Communications
in Computer and Information Science 116

Luciano da F. Costa   Alexandre Evsukoff
Giuseppe Mangioni   Ronaldo Menezes (Eds.)

Complex Networks

Second International Workshop, CompleNet 2010
Rio de Janeiro, Brazil, October 13-15, 2010
Revised Selected Papers

 Springer

Volume Editors

Luciano da F. Costa
University of São Paulo
Institute of Physics at São Carlos
P.O.Box 369, São Carlos, 13560-970 São Paulo, Brazil
E-mail: luciano@if.sc.usp.br

Alexandre Evsukoff
COPPE/Federal University of Rio de Janeiro (UFRJ)
P.O. Box 68506, 21941-972 Rio de Janeiro RJ, Brazil
E-mail: alexandre.evsukoff@coc.ufrj.br

Giuseppe Mangioni
University of Catania
Dipartimento di Ingegneria Elettrica, Elettronica e Informatica (DIEEI)
Viale A. Doria, 6, 95125 Catania, Italy
E-mail: giuseppe.mangioni@dieei.unict.it

Ronaldo Menezes
Florida Institute of Technology
Computer Sciences
150 W. University Blvd, Melbourne, FL, 32901, USA
E-mail: rmenezes@cs.fit.edu

ISSN 1865-0929   e-ISSN 1865-0937
ISBN 978-3-642-25500-7   e-ISBN 978-3-642-25501-4
DOI 10.1007/978-3-642-25501-4
Springer Heidelberg Dordrecht London New York

Library of Congress Control Number: 2011941114

CR Subject Classification (1998): C.2, H.4, H.3, D.2, I.2, C.2.4, I.4

Typesetting: Camera-ready by author, data conversion by Scientific Publishing Services, Chennai, India

Printed on acid-free paper

Springer is part of Springer Science+Business Media (www.springer.com)

Preface

The International Workshop on Complex Networks—CompleNet
(www.complenet.org)—was initially proposed in 2008 with the first workshop
taking place in 2009. The initiative was the result of efforts from researchers
from the Bio-Complex Laboratory in the Department of Computer Sciences at
Florida Institute of Technology, USA, and the Dipartimento di Ingegneria Elettrica, Elettronica e Informatica, Università di Catania, Italy. CompleNet aims
at bringing together researchers and practitioners working on areas related to
complex networks. In the past two decades we have been witnessing an exponential increase in the number of publications in this field. From biological systems
to computer science, from economic to social systems, complex networks are becoming pervasive in many fields of science. It is this interdisciplinary nature of
complex networks that CompleNet aims at addressing. CompleNet 2010 was the
second event in the series and was hosted by the Núcleo de Transferência de
Technologia at the Federal University of Rio de Janeiro (UFRJ) during October
13–15, 2010.

This book includes the peer-reviewed list of works presented at CompleNet
2010. Submissions were accepted either as a paper or as an abstract (presentation only). We received 48 submissions from 18 countries. Each submission was
reviewed by at least three members of the Program Committee. Acceptance was
judged based on the relevance to the symposium themes, clarity of presentation, originality and accuracy of results and proposed solutions. After the review
process, eight papers and nine short papers were selected for presentation. We
also invited 24 abstracts for presentation only. In this volume we have included
the 21 papers and short papers plus a very selected number of abstracts. The
authors of abstracts were invited to submit a paper after their presentation at
CompleNet and the papers went through a second round of peer revision.

The 21 contributions in this book address many topics related to complex networks including: community structure, network metrics, network models, effect
of topology to epidemics, algorithms to classify networks, self-organized algorithms applied to complex networks, as well as many applications of complex
networks in biology, image analysis, software development, traffic congestion,
language and speech, sensor networks, and synchronization.

We would like to thank to the Program Committee members for their work
in promoting the event and in refereeing submissions. We deeply appreciate the
efforts of our keynote speakers: M. Madan Babu (University of Cambridge, UK)
and Eugene Stanley (Boston University, USA); their presentation is one of the
reasons CompleNet 2010 was such a success. We are grateful to our invited speakers who enriched CompleNet 2010 with their presentations and insights in the
field of Complex Networks (in alphabetical order): Cesar Hidalgo (MIT, USA),
Jure Lescovec (Stanford University, USA), Márcio A. de Menezes (Universidade

Federal Fluminense, Brazil), Maximilian Schich (Northeastern University, USA), and Rita Zorzenon (Universidade Federal de Pernambuco, Brazil).

Special thanks also go to Nelson Ebecken and Beatriz Lima from Universidade Federal do Rio de Janeiro and all the members from the Núcleo de Transferência de Technologia for their help in organizing CompleNet 2010. The next edition of CompleNet will be hosted by the Florida Institute of Technology, USA, in the beginning of 2012.

October 2010   Luciano Costa
                                                   Alexandre Evsukoff
                                                   Giuseppe Mangioni
                                                   Ronaldo Menezes

Conference Organization

Steering Committee

Giuseppe Mangioni	University of Catania, Italy
Ronaldo Menezes	Florida Institute of Technology, USA
Vincenzo Nicosia	University of Catania, Italy

Workshop Chair

Alexandre Evsukoff	Fedaral University of Rio de Janeiro, Brazil

Program Committee Chairs

Luciano Costa	University of São Paulo, Brazil
Giuseppe Mangioni	University of Catania, Italy

Program Committee

Yong-Yeol Ahn	Northeastern University, USA
Roberto Andrade	Universidade Federal da Bahia, Brazil
Alex Arenas	Universitat Rovira i Virgili, Tarragona, Spain
Madan Babu	University of Cambridge, UK
Rodolfo Baggio	Bocconi University, Milan, Italy
Alain Barrat	CPT Marseille, France
Marc Barthelemy	Institut de Physique Théorique, CEA, France
Stefano Battiston	ETH, Zurich, Switzerland
Ana Bazzan	Federal University of Rio Grande do Sul, Brazil
Ginestra Bianconi	Northeastern University Boston, USA
Haluk Bingol	Bogazici University, Istanbul, Turkey
Stefan Bornholdt	University of Bremen, Germany
Dan Braha	University of Massachusetts, Dartmouth, USA
Ulrik Brandes	University of Konstanz, Germany
Guido Caldarelli	INFM, Rome, Italy
Vincenza Carchiolo	University of Catania, Italy
Claudio Castellano	INFM, Rome, Italy
Aaron Clauset	Santa Fe Institute, USA
Regino Criado	Universidad Rey Juan Carlos, Madrid, Spain
Giorgio Fagiolo	Sant'Anna School of Advanced Studies, Pisa, Italy

Santo Fortunato	Complex Networks Lagrange Lab (ISI), Turin, Italy
Kwang-Il Goh	Korea University, Seoul, Republic of Korea
Marta González	MIT, USA
Steve Gregory	University of Bristol, UK
Roger Guimera	Northwestern University, Evanston, USA
Jens Gustedt	INRIA Nancy - Grand Est, France
Cesar Hidalgo	Harvard University, USA
Claus Hilgetag	Jacobs University, Bremen, Germany
Kimmo Kaski	Helsinki University of Technology, Finland
János Kertész	Budapest University of Technology and Economics, Hungary
Renaud Lambiotte	Imperial College, London, UK
Matthieu Latapy	LIP6 - CNRS et Université Paris 6, Paris, France
Anna Lawniczak	University of Guelph, Canada
Sune Lehmann	Northeastern University, USA
Alessandro Longheu	University of Catania, Italy
Jose Mendes	University of Aveiro, Portugal
Adilson E. Motter	Northwestern University, Evanston, USA
Osvaldo Novais de Oliveira	IFSC, University of Sao Paulo, Brazil
Eraldo Ribeiro	Florida Institute of Technology, USA
Martin Rosvall	Umea University, Sweden
M. Ángeles Serrano	Universitat de Barcelona, Spain
Filippo Simini	University of Padova, Italy
Igor Sokolov	Humboldt University, Berlin, Germany
Bosiljka Tadic	Jozef Stefan Institute, Ljubljana, Slovenia
Sergei N. Taraskin	University of Cambridge, UK
Andres Upegui	HEIG-VD, Switzerland
Soon-Hyung Yook	Kyung Hee University, Seoul, Republic of Korea
Rita M. Zorzenon dos Santos	Universidade Federal de Pernambuco, Brazil

Local Organizing Committee

Alexandre Evsukoff	Fedaral University of Rio de Janeiro
Nelson Ebecken	Federal University of Rio de Janeiro
Beatriz Lima	Federal University of Rio de Janeiro

Table of Contents

Network Modeling

Applications

Network Dynamics

Community Structure

On the Average Path Length of Deterministic and Stochastics Recursive Networks⋆

Philippe J. Giabbanelli, Dorian Mazauric, and Jean-Claude Bermond

Mascotte, INRIA, I3S(CNRS,UNS), Sophia Antipolis, France
{Philippe.Giabbanelli,Dorian.Mazauric,
Jean-Claude.Bermond}@sophia.inria.fr

Abstract. The average shortest path distance ℓ between all pairs of nodes in real-world networks tends to be small compared to the number of nodes. Providing a closed-form formula for ℓ remains challenging in several network models, as shown by recent papers dedicated to this sole topic. For example, Zhang *et al.* proposed the deterministic model ZRG and studied an upper bound on ℓ. In this paper, we use graph-theoretic techniques to establish a closed-form formula for ℓ in ZRG. Our proof is of particular interest for other network models relying on similar recursive structures, as found in fractal models. We extend our approach to a stochastic version of ZRG in which layers of triangles are added with probability p. We find a first-order phase transition at the critical probability $p_c = 0.5$, from which the expected number of nodes becomes infinite whereas expected distances remain finite. We show that if triangles are added independently instead of being constrained in a layer, the first-order phase transition holds for the very same critical probability. Thus, we provide an insight showing that models can be equivalent, regardless of whether edges are added with grouping constraints. Our detailed computations also provide thorough practical cases for readers unfamiliar with graph-theoretic and probabilistic techniques.

1 Introduction

The last decade has witnessed the emergence of a new research field coined as "Network Science". Amongst well-known contributions of this field, it was found that the average distance ℓ in a myriad of real-world networks was small compared to the number of nodes (*e.g.*, in the order of the logarithm of the number of nodes). Numerous models were proposed for networks with small average distance [1,2] such as the static Watts-Strogatz model, in which a small percentage of edges is changed in a low-dimensional lattice [3], or dynamic models in which ℓ becomes small as nodes are added to the network [4]. However, proving a closed form formula for ℓ can be a challenging task in a model, and thus this remains a current research problem with papers devoted to this sole task [5]. In this paper, we prove a closed form formula for a recently proposed model, in which the authors showed an upper bound on ℓ [6]. While the model presented in [6] is deterministic, a stochastic version was also studied for which the authors

---

⋆ Research funded by the EULER project and *région PACA*.

L. da F. Costa et al. (Eds.): CompleNet 2010, CCIS 116, pp. 1–12, 2011.
© Springer-Verlag Berlin Heidelberg 2011

approximated an upper bound on ℓ [8]. Thus, we present two stochastic versions and we rigorously characterize their behaviour using both upper and lower bounds on ℓ, and studying the ratio with the number of nodes.

The paper starts by establishing the notation, then each of the three Sections focusses on a model. Firstly, we consider the model as defined in [6]: we prove a closed-form formula for the average distance ℓ, and characterize the ratio between the number of nodes and ℓ. Secondly, we propose a version of the model in which edges and nodes are randomly added but in specific groups. In this version, we establish bounds on the expected value of ℓ and we provide a closed-form formula for the expected number of nodes. While the former is always finite, the latter becomes infinite from a critical probability $p_c = 0.5$, thus the ratio between ℓ and the number of nodes can be arbitrarily large. However, the infinite value of the expected number of nodes results from a few very large instances, and thus does not represent well the trend expressed by most instances for $p \geq p_c$. Consequently, we also study the ratio between the number of nodes and ℓ by considering all instances but very large ones. Thirdly, we study the number of nodes and ℓ in a stochastic version that does not impose specific groups, similarly to [8]. We show that this version also has a finite expected value for ℓ, and an infinite expected number of nodes from $p = p_c$.

2   Notation

We denote by ZRG_t the undirected graph defined by Zhang, Rong and Guo, obtained at step t [6]. It starts with ZRG_0 being a cycle of three nodes, and "ZRG_t is obtained by ZRG_{t-1} by adding for each edge created at step $t-1$ a new node and attaching it to both end nodes of the edge" [6]. The process is illustrated by Figure 1(a). We propose two probabilistic versions of ZRG. In the first one, each of the three original edges constitutes a *branch*. At each time step, a node is added *for all* active edges of a branch with independent and identical (*iid*) probability p. If a branch does not grow at a given time step, then it will not grow anymore. We denote this model by $BZRG_p$, for the probabilistic *branch* version of ZRG with probability p. Note that while the probability p is applied at each time step, the resulting graph is not limited by a number of time steps as in ZRG_t: instead, the graph grows as long as there is at least a branch for which the outcome of the stochastic process is to grow, thus there exist arbitrarily large instances. The process is illustrated in Figure 1(b). Finally, the second stochastic version differs from $BZRG_p$ by adding a node with probability p *for each* active edge. In other words, this version does not impose to grow all the 'layer' at once, but applies the probability edge by edge. We denote the last version by $EZRG_p$ for the probabilistic *edge* version of ZRG with probability p.

In this paper, we are primarily interested in the average distance. For a connected graph G having a set of nodes $V(G)$, its average distance is defined by $\ell(G) = \frac{\sum_{u \in V(G)} \sum_{v \in V(G)} d(u,v)}{|V(G)| * (|V(G)| - 1)}$, where $d(u, v)$ is the length of a shortest path between u and v. In a graph with N nodes, ℓ is said to be *small* when proportional to $\ln(N)$, and *ultrasmall* when proportional to $\ln(\ln(N))$ [9].

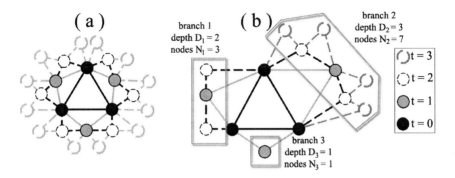

Fig. 1. The graph ZRG_0 is a triangle, or cycle of 3 nodes. At each time step, to each edge added at the previous time step, we add a node connected to the endpoints of the edge. This can be seen as adding a triangle to each outer edge. The process is depicted step by step up to ZRG_3 (a). A possible instance of $BZRG_p$ illustrates the depth and number of nodes in each of the three probabilistic branches (b). The graph grew for 3 time steps, at the end of which the outcome of the last active branch was not to grow.

3   Deterministic Version

In this Section, we consider the version introduced by [6] and defined in the previous Section. We denote by V_t the vertices of ZRG_t, and A_t the vertices added at step t. We established in [7] that $|A_t| = 3 * 2^{t-1}$ for $t \geq 1$.

By construction, a node $u \in A_t$ is connected to the two endpoints of a formerly active edge. One endpoint was created at the step immediately before (*i.e.*, $t-1$), and we call it the *mother* $m(u) \in A_{t-1}$, while the other endpoint was created at an earlier time, and we call it the *father* $f(u) \in V_{t-2}$. A node having same mother as u is called its uterine brother and denoted as $b(u)$. Furthermore, we observe that each node $v \in V_{t-1}$ is connected to two nodes of A_t. This is proved by induction: it holds for $t = 1$ (see Figure 2(a)), we assume it holds up to $t - 1$ and we show in Figure 2(b-c) that it follows for t. Since each node in A_t is connected to two nodes in V_{t-1}, and each node in V_{t-1} is connected to two nodes in A_t, the graph has a bipartite structure used in our proof.

We now turn to the computation of $\ell(ZRG_t)$. We denote by $d(X,Y) = \sum_{u \in X} \sum_{v \in Y} d(u,v)$ the sum of distances from all nodes in X to all nodes in Y. Theorem 1 establishes the value of $g(t) = d(V_t, V_t)$, from which we will establish the average distance using $\ell(ZRG_t) = \frac{g(t)}{|V(ZRG_t)|*(|V(ZRG_t)|-1)}$. The following Theorem was announced in [7] with a different proof only sketched there.

Theorem 1. $g(t) = 4^t(6t + 3) + 2 * 3^t$

Proof. By definition, $V_t = V_{t-1} \cup A_t$. Thus, $d(V_t, V_t) = d(V_{t-1}, V_{t-1}) + d(A_t, V_{t-1}) + d(V_{t-1}, A_t) + d(A_t, A_t)$. Since the underlying graph is undirected, $d(A_t, V_{t-1}) = d(V_{t-1}, A_t)$ hence

$$g(t) = g(t-1) + 2d(A_t, V_{t-1}) + d(A_t, A_t), t \geq 2 \qquad (1)$$

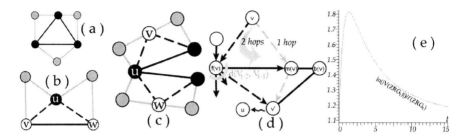

Fig. 2. In ZRG_1, each of the black initial nodes is connected to two grey added nodes (a). We assume that $u \in A_{t-1}$, thus it stems from an edge (v, w). As the edges (u, v) and (u, w) are active, u will also be connected to two nodes in A_t (b). We assume that $u \in V_{t-2}$: thus, it is connected to two (children) nodes $v, w \in A_{t-1}$. The edges (u, v) and (u, w) being active, u will also be connected to the two nodes they generate at the next step, belonging to A_t (c). Shortest paths to compute $d(A_t, V_{t-1})$ (d). The average distance in ZRG_t is close to the logarithm of the graph's size (e).

In the following, we consider that a shortest path from $v \in A_t$ always goes through $f(v)$, unless the target is the brother $b(v)$ or the mother $m(v)$ in which case correction factors are applied. Suppose that we instead go through $m(v)$ to reach some node u: since $m(v)$ is only connected to $b(v)$, $f(v)$ and some node v' (see Figure 2(d)) then the route continues through v'. However, the path $v, m(v), v'$ can be replaced by $v, f(v), v'$ without changing the length.

We compute $d(A_t, V_{t-1})$ for $t \geq 2$. Since we always go through $f(v)$, we use a path of length 2 in order to go from v to $m(v)$ whereas it takes 1 using the direct link. Thus, we have to decrease the distance by 1 for each $v \in A_t$, hence a correcting factor $-|A_t|$. We observe that each node in V_{t-2} is the father of two nodes in A_t, hence routing through the father costs $2d(V_{t-2}, V_{t-1})$ to which we add the number of times we use the edge from v to the father. As each $v \in A_t$ goes to each $w \in V_{t-1}$, the total cost is

$$d(A_t, V_{t-1}) = 2d(V_{t-2}, V_{t-1}) + |A_t||V_{t-1}| - |A_t| \qquad (2)$$

We have that $2d(V_{t-2}, V_{t-1}) = 2d(V_{t-2}, V_{t-2}) + 2d(V_{t-2}, A_{t-1}) = 2g(t-2) + 2d(V_{t-2}, A_{t-1})$. Furthermore, using Eq. 1 we obtain $g(t-1) = g(t-2) + 2d(A_{t-1}, V_{t-2}) + d(A_{t-1}, A_{t-1}) \Leftrightarrow 2d(A_{t-1}, V_{t-2}) = g(t-1) - g(t-2) - d(A_{t-1}, A_{t-1})$. Substituting these equations with Eq. 2, it follows that

$$d(A_t, V_{t-1}) = g(t-1) + g(t-2) - d(A_{t-1}, A_{t-1}) + |A_t||V_{t-1}| - |A_t| \quad (3)$$

We compute $d(A_t, A_t)$, for $t \geq 2$. In order to go from v to its uterine brother $b(v) \in A_t$, it takes 2 hops through their shared mother, whereas it takes 3 hops through the father. Thus, we have a correction of 1 for $|A_t|$ nodes. The path from a v to a w is used four times, since $f(v)$ has two children in A_t and so does $f(w)$. Finally, we add 2 for the cost of going from a node to its father at both ends of the path, and we have $|A_t|(|A_t| - 1)$ such paths. Thus, the overall cost is

$$d(A_t, A_t) = 4g(t-2) + 2|A_t|(|A_t| - 1) - |A_t| \qquad (4)$$

We combine. Given that $|A_t| = |V_{t-1}|$, we substitute Eq. 3 into Eq. 1 hence

$$g(t) = 3g(t-1) + 2g(t-2) + d(A_t, A_t) - 2d(A_{t-1}, A_{t-1}) + 2|A_t|^2 - 2|A_t| \quad (5)$$

From Eq. 4, for $t \geq 3$ we obtain $d(A_{t-1}, A_{t-1}) = 4g(t-3) + 2|A_{t-1}|^2 - 3|A_{t-1}|$. Given that $|A_t| = 2|A_{t-1}|$, we substitute $d(A_{t-1}, A_{t-1})$ and Eq. 4 in Eq. 5:

$$g(t) = 3g(t-1) + 6g(t-2) - 8g(t-3) + 3|A_t|^2 - 2|A_t|, t \geq 3 \quad (6)$$

We manually count that $f(0) = 6$, $f(1) = 42$ and $f(2) = 252$. Thus the equation can be solved into $g(t) = 4^t(6t+3) + 3*2^t$ using standard software.

The Corollary follows from Theorem 1 by using the fact that $|V(ZRG_t)| = 3*2^t$ [6].

Corollary 1. $\ell(ZRG_t) = \frac{4^t(6t+3)+3*2^t}{3*2^t(3*2^t-1)} = \frac{t2^{t+1}+2^t+1}{3*2^t-1}$.

Using this corollary, we obtain $\lim_{t\to\infty} \frac{\ln(|V(ZRG_t)|)}{\ell(ZRG_t)} = \frac{3*\ln(2)}{2} \approx 1.03$ and $\lim_{t\to\infty} \frac{\ln(\ln(|V(ZRG_t)|))}{\ell(ZRG_t)} \approx 0$. Thus, the average size is almost exactly $\ln(|V(G)|)$ for large t. Since the size of the graph is exponential in t, it is important that the graphs obtained for small values of t have a similar ratio, which is confirmed by the behaviour illustrated in Figure 2(e).

4   Stochastic Branch Version

As in the previous Section, we are interested in the ratio between the number of nodes and the average path length. In this Section, our approach is in three steps. Firstly, we establish bounds on the *depth* of branches, defined as the number of times that a branch has grown. Secondly, we study the number of nodes. We find that the number of nodes undergoes a first-order phase transition at the critical probability $p_c = 0.5$: for $p < 0.5$, there is a finite number of nodes, whereas for $p \geq 0.5$ this number becomes infinite. Since in the latter the expected depth of branches is bounded by finite numbers, the expected graphs have an arbitrarily small average distance compared to the number of nodes. However, the expected number of nodes only provides a mean-field behaviour that can may lack representativeness due to a few very large instances. Thus, we conclude by investigating the behavior of instances of bounded depth.

4.1   Depth of Branches

To fully characterize the depth of branches, we are interested in their expected depth for the standard case as well as the two extremal cases consisting of the *deepest* and *shallowest* branches. In other words, we study the mean-field behaviour and we provide a lower and an upper bound. We start by introducing our notation for the depth in Definition 1. We start by establishing the expected depth of a branch in Theorem 2, then we turn to the expected shallowest depth, and we conclude in Theorem 4 showing that the expected deepest depth of a branch is finite.

Definition 1. *We denote D_1, D_2, D_3 the depth of the three branches. The depth of the deepest branch is $D_{max} = max(D_1, D_2, D_3)$ and the depth of the shallowest branch is $D_{min} = min(D_1, D_2, D_3)$.*

Theorem 2. *The expected depth of a branch is $\mathbb{E}(D_i) = \frac{p}{1-p}$, $i \in \{1, 2, 3\}$.*

Proof. The probability $P(D_i = k)$ that a branch grows to depth k is the probability p^k of successfully growing k times, and the probability not to grow once (*i.e.*, to stop at depth $k + 1$). Thus, $P(D_i = k) = \underbrace{p \cdots \cdots p}_{k} \cdot (1 - p) = p^k(1 - p)$. Since the expected value of a discrete random variable D_i is given by $\mathbb{E}(D_i) = \sum_{k=0}^{\infty}(kP(D_i = k))$, it follows that $\mathbb{E}(D_i) = \sum_{k=0}^{\infty}(kp^k(1 - p)) = p(1 - p)\sum_{k=0}^{\infty}(kp^{k-1})$. Since $\frac{dp^k}{dp} = kp^{k-1}$, we further simplify into $\mathbb{E}(D_i) = p(1 - p)\sum_{k=0}^{\infty}(\frac{dp^k}{dp})$. As the sum of a derivative is equal to the derivative of the sum, it follows that $\mathbb{E}(D_i) = p(1 - p)\frac{d\sum_{k=0}^{\infty}p^k}{dp}$. We note that $\sum_{k=0}^{\infty}p^k$ is an infinite geometric sum, hence

$$\mathbb{E}(D_i) = p(1 - p)\frac{d}{dp}\frac{1}{1-p} = \frac{p(1-p)}{(1-p)^2} = \frac{p}{1-p}$$

Theorem 3. $\mathbb{E}(D_{min}) = \frac{p^3}{1-p^3}$

Proof. The probability of the shallowest depth to be at least k knowing that the probability p applies iid to each branch is $P(D_{min} \geq k) = P(D_1 \geq k)P(D_2 \geq k)P(D_3 \geq k) = p^{3k}$. By definition, $P(D_{min} = k) = P(D_{min} \geq k) - P(D_{min} \geq k + 1)$, thus $P(D_{min} = k) = p^{3k} - p^{3(k+1)}$. This probability is plugged into the definition of the expected value as in Theorem 2 hence

$$\mathbb{E}(D_{min}) = \sum_{k=0}^{\infty}(kP(D_{min} = k)) = \sum_{k=0}^{\infty}(k(p^{3k} - p^{3(k+1)})) = -\frac{p^3}{p^3-1}$$

Theorem 4. $\mathbb{E}(D_{max}) = \frac{p(p^3+4p^2+3p+3)}{(1-p)(p^2+p+1)(p+1)}$.

Proof. By construction, the deepest branch does not exceed k iff none of the branches has a depth exceeding k. Since the probability p applies iid to all three branches, we have $P(D_{max} \leq k) = P(D_1 \leq k)P(D_2 \leq k)P(D_3 \leq k)$. Furthermore, a branch is strictly deeper than k if it successfully reaches depth $k + 1$. Thus, $P(D_i > k) = \underbrace{p \cdots \cdots p}_{k+1} = p^{k+1}$, $i \in \{1, 2, 3\}$. By algebraic simplification, we have $P(D_{max} \leq k) = (1 - P(D_1 > k))(1 - P(D_2 > k))(1 - P(D_3 > k)) = (1 - p^{k+1})^3$. By definition, $P(D_{max} = k) = P(D \leq k) - P(D \leq k - 1) = (1 - p^{k+1})^3 - (1 - p^k)^3$. Given that $\mathbb{E}(D_{max}) = \sum_{k=0}^{\infty}(kP(D_{max} = k))$, we replace the expression of $P(D_{max} = k)$ to obtain $\mathbb{E}(D_{max}) = \sum_{k=0}^{\infty}(k((1 - p^{k+1})^3 - (1 - p^k)^3))$. The final expression results from algebraic simplification using standard software.

4.2   Average Number of Nodes

We introduce our notation in Definition 2, and Theorem 5 provides a closed-form of the expected number of nodes.

Definition 2. *We denote by N_1, N_2, and N_3 the number of nodes in the three branches. Since we start from a cycle with three nodes, the total number of nodes is $N = N_1 + N_2 + N_3 + 3$.*

Theorem 5. *For $p < \frac{1}{2}$, the expected number of nodes is $\mathbb{E}(N) = \frac{3(1-p)}{1-2p}$.*

Proof. First, we focus on the expected number of nodes in a branch. As the probability p applies iid to all three branches, we select the first branch without loss of generality. By construction, the total number of nodes N_1 in the branch 1 at depth $D_1 = k \geq 0$ is $N_1 = 2^k - 1 = \sum_{i=1}^{k} 2^{i-1}$. Thus, the expected value of the random variable N_1 is given by $\mathbb{E}(N_1) = \sum_{k=0}^{\infty}((2^k - 1)P(D_1 = k))$. As shown in Theorem 2, $P(D_1 = k) = p^k(1 - p)$. We replace it in the equation to obtain $\mathbb{E}(N_1) = \sum_{k=0}^{\infty}((2^k - 1)p^k(1-p))$. After expanding the equation, we have $\mathbb{E}(N_1) = \sum_{k=0}^{\infty}(2^k p^k(1-p) - p^k(1-p)) = (1-p)\sum_{k=0}^{\infty}((2p)^k) - (1-p)\sum_{k=0}^{\infty}(p^k)$. As in Theorem 2, we have $\sum_{k=0}^{\infty}(p^k) = \frac{1}{1-p}$ thus the second term simplifies and yields $\mathbb{E}(N_1) = (1 - p)\sum_{k=0}^{\infty}((2p)^k) - 1$. It is well known that series of the type $\sum_{k=0}^{\infty}(x^k)$ diverges to infinity for $x \geq 1$. Thus, our series diverges for $2p \geq 1 \Leftrightarrow p \geq \frac{1}{2}$. In other words, this results only holds for $0 \leq p < \frac{1}{2}$. The infinite geometric sum $\sum_{k=0}^{\infty}((2p)^k)$ can be simplified in $\frac{1}{1-2p}$ hence $\mathbb{E}(N_1) = \frac{1-p}{1-2p} - 1 = \frac{1-p-1+2p}{1-2p} = \frac{p}{1-2p}$. The probability p applies iid to all three branches hence $\mathbb{E}(N_1) = \mathbb{E}(N_2) = \mathbb{E}(N_3)$. Thus, by Definition 2, the expected number of nodes in the overall graph is given by $\mathbb{E}(N) = 3 + 3\mathbb{E}(N_1) = 3 + \frac{3p}{1-2p} = \frac{3(1-p)}{1-2p}$. □

Theorem 5 proved that the average number of nodes diverges to infinity at the critical probability $p_c = 0.5$. This may appear to be a discrepancy with Theorem 4 stating that the expected depth of the deepest branch is finitely bounded. For the sake of clarity, we provide an intuition and an example on this point. First, we note that the *expected* deepest depth and the *expected* number of nodes have different growth rates. Indeed, even if graphs with very deep depth scarcely occur at $p = 0.5$, their impact on the expected number of nodes is tremendous since the number of nodes grows *exponentially* with the depth. On the other hand, the impact of such graphs on the expected deepest depth is only linear. To illustrate different growth rates with a known network, consider the complete graph K_n, in which each of the $n \geq 1$ nodes is connected to all others. In K_n, the number of nodes grows linearly whereas the distance is constant. Thus, the distance between two nodes is 1 even with an infinite number of nodes.

In a nutshell, the expected number of nodes for $p \geq 0.5$ may not be representative of most instances due to the large impact of very deep graphs. Thus, it remains of interest to investigate the number of nodes for graphs with bounded depths. This is established in Theorem 6, in which we consider the $q\%$ possible instances with smallest depth. By lowering the impact of the very deep graphs, this theorem consistutes a lower bound that better describes practical cases.

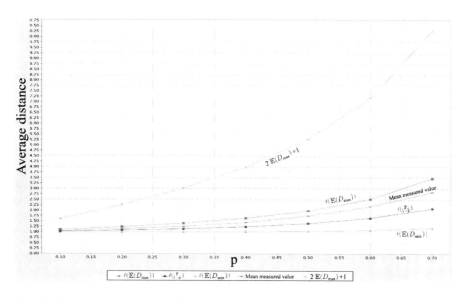

Fig. 3. The measured average distance compared with three bounds and an estimate, proved or conjectures. The simulations valide the conjectures. *Color online.*

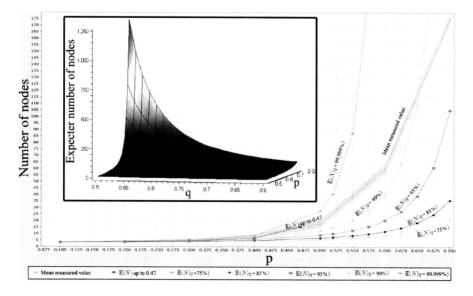

Fig. 4. Up to $p = 0.5$ (excluded), the expected number of nodes is finite. From $p = 0.5$, the expected number of nodes is infinite due to very large instances, thus we provide a finite estimate by considering the $q\%$ smallest instances, from $q = 75\%$ to $q = 100 - 10^{-3}\%$. Simulations were limited to the graphs that fitted in memory thus the mean measured value represents only an estimate based on small graphs for values beyond $p = 0.5$. *Color online.*

Lemma 1. $\mathbb{E}(N|D_{max} \leq K) = -\frac{3(1-p)(2^{K+1}p^{K+1}-1)}{(2p-1)(1-p^{K+1})}$

Proof. The expected number of nodes is adapted from Theorem 5 by the following substitutions: $\mathbb{E} \underbrace{(N_1)}_{(N_1|D_{max}\leq K)} = \sum_{k=0}^{\infty}((2^k - 1)\underbrace{P(D_1 = k)}_{P(D_1=k|D_{max}\leq K)})$. Since the branches are independent, we get $P(D_1 = k|D_{max} \leq K) = \frac{P(D_1=k \cap D_1 \leq K)}{P(D_1 \leq K)} = \frac{P(D_1=k)}{P(D_1 \leq K)}$ if $k \leq K$ and 0 otherwise. Thus $\mathbb{E}(N_1|D_{max} \leq K) = \sum_{k=0}^{K}((2^k - 1)\frac{P(D_1=k)}{P(D_1 \leq K)})$. We showed in the proof of Theorem 4 that $P(D_1 \leq K) = 1-p^{K+1}$, and we showed in the proof of Theorem 2 that $P(D_1 = k) = p^k(1 - p)$. By substituting these results, and using from the previous Theorem that $\mathbb{E}(N) = 3 + 3\mathbb{E}(N_1)$, it follows that

$$\mathbb{E}(N|D_{max} \leq K) = 3 + 3\sum_{k=0}^{K}(\frac{(2^k-1)p^k(1-p)}{1-p^{K+1}})$$

The closed form formula follows by algebraic simplification.

Theorem 6. *By abuse of notation, we denote by $\mathbb{E}(N|q)$ the expected number of nodes for the $q\%$ of instances of $BZRG_p$ with smallest depth. We have*

$$\mathbb{E}(N|q) = -\frac{3(1-p)((1-\sqrt[3]{q})^{\frac{\ln(2)+\ln(p)}{\ln(p)}}-1)}{\sqrt[3]{q}(2p-1)}$$

Proof. Theorem 4 proved that the expected deepest depth of a branch was at most K with probability $(1 - p^{K+1})^3$. Thus, if we want this probability to be q, we have to consider branches whose depth K is at most:

$$(1 - p^{K+1})^3 = q \Leftrightarrow K = log_p(1 - \sqrt[3]{q}) - 1$$

The Theorem follows by replacing K in Lemma 1 with the value above.

The effect of a few percentages of difference in q is shown in Figure 4 together with the results from Theorem 5 and 6. In the inset of Figure 4, we show that the number of nodes grows sharply with q.

4.3   Average Path Length

We conducted experiments in order to ensure the veracity of the results presented in the two previous Sections, and to compare them with devised bounds. For values of p from 0.1 to 0.7 by steps of 0.1, we measured the average distance of the resulting graphs, obtained as the average over 1000 instances. In Figure 3, we plot it against four bounds and an estimated mean:

- *Proven bound.* Theorem 4 established the expected deepest depth. At any time, the graph has three branches, and we can only go from one branch to another through the basic cycle. Thus, the expected maximum distance between any two points in the graph consists of going from the most remote node of two branches to the cycle, and going through the cycle. As a node is at most at distance $\mathbb{E}(D_{max})$ from the cycle and we can go from any node of the cycle to another in one hop, the expected maximum distance is

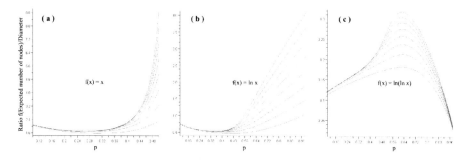

Fig. 5. We measured the ratio between $\mathbb{E}(N|100 - 10^x)$ (a), $\ln(\mathbb{E}(N|100 - 10^x))$ (b), $\ln(\ln(\mathbb{E}(N|100 - 10^x)))$ (c) and the diameter $2\mathbb{E}(D_{max}) + 1$ for x going from 0 (bottom curve) to 7 (top curve). We determined a critical probability $p_c = 0.5$ at which the regime changes, and this is confirmed by these measures showing that the average distance goes from linear in the number of nodes (a) for $p < p_c$ to small in the number of nodes (b) for $p \geq p_c$.

$2\mathbb{E}(D_{max}) + 1$. Since this is the *maximum* distance, we use it as a proven upper bound on the *average* distance.

- *Conjectured bounds.*Corollary 1 proved the average distance $\ell(ZRG_t)$ when *all* branches have depth t. Thus, we conjecture that a lower and upper bound can be obtained by considering the graphs with shallowest (Theorem 3) and deepest (Theorem 4) depths respectively. This is confirmed by the simulations.
- *Conjectured mean.* Similarly to the conjecture bounds, we have proven the expected depth $\mathbb{E}(D) = \frac{p}{1-p}$ of a branch in Theorem 2, and the simulation confirms that $\ell(\frac{p}{1-p})$ constitute an estimate of the average distance.

As we previously did for the deterministic version, we now investigate whether the average distance $\ell(BZRG_p)$ can be deemed small compared to the number of nodes $|V|$. As explained in the previous Section, we proved a (first-order) phase transition at the critical probability $p_c = 0.5$. The behaviour of the graph can be characterized using the ratios displayed in Figure 5: for $p \ll p_c$, we observe an average distance proportional to the number of nodes, and for $p > p_c - \epsilon$ the average distance is proportional to the logarithm of the number of nodes which is deemed small. The ratio in Figure 5(c) is too low, and tends to 0 for a large probability p, thus the graph cannot be considered ultra-small. The separation at $p_c - \epsilon$ can also be understood from a theoretical perspective. For $p \geq p_c$, we proved that $\ell(BZRG_p)$ can be arbitrary small compared to $|V|$ since $\ell(BZRG_p)$ is finite whereas $|V|$ is infinite. When $p = 0.5 - \varepsilon$, the average distance is bounded by a finite number: by Theorem 2 we have that the expected depth of a branch is $\mathbb{E}(D_i) < 1$ and, using the aforementioned argument regarding the maximum distance, this entails $\ell(BZRG_p) < 2 * 1 + 1 = 3$. Furthermore, as stated in the proof of Theorem 6, the expected number of nodes in a branch is $\mathbb{E}(N_i) = \frac{0.5 - \varepsilon}{\varepsilon}$ which can thus be arbitrarily large. Thus, the behaviour for $p \geq p_c$ is also expected to hold in a small neighborhood of the critical probability.

5   Stochastic Edge Version

In order to show a broad variety of approaches, we prove the number of nodes and the average path length of $EZRG_p$ using different tools from the previous Section. Theorem 7 establishes the number of nodes in the graph.

Theorem 7. *For $p < \frac{1}{2}$, the expected number of nodes is $\mathbb{E}(N) = 3 + \frac{3p}{1-2p}$.*

Proof. We consider the dual of the graph, which we define using a visual example in Figure 6. The dual is a binary tree: for an edge, a triangle is added (root of the tree) with probability p, to which two triangles are added iid at the next step (left and right branches of the tree) with probability p, and so on. Since one node is added to the tree when a node is added to the original graph, studying the number of nodes in $EZRG_p$ is equivalent to studying the number of nodes in the tree. We denote the latter by $t(p)$. The number of nodes starting from any edge is the same, and we denote it by N. Thus, N corresponds to the probability of starting a tree (*i.e.*, adding a first node that will be the root) multipled by the number of nodes in the tree: $N = pt(p)$. Once the tree has been started, the number of nodes corresponds to the sum of the root and the number of nodes in the two branches, hence $t(p) = 2pt(p) + 1$. Note that there is a solution if and only if $p < \frac{1}{2}$, and otherwise the number of nodes is infinite. By arithmetic simplification, we obtain $t(p)(1 - 2p) = 1 \Leftrightarrow t(p) = \frac{1}{1-2p}$. Thus, $N = \frac{p}{1-2p}$. Since the graph starts as a cycle of length three, the number of counts corresponds to the three original nodes, to which we add the number of nodes starting in each of the three trees, thus $\mathbb{E}(N) = 3 + \frac{3p}{1-2p} = \frac{3(1-p)}{1-2p}$. □

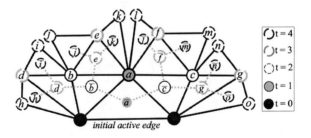

Fig. 6. Nodes linked by black edges correspond to four successive growths from an initial active edge. When a node x is added, it creates a triangle, to which we associate a node $\bar{x}$. If two triangles share an edge, their nodes $\bar{x}_1$ and $\bar{x}_2$ are connected by a grey dashed edge. The graph formed by the nodes associated to triangles is called *dual*.

A proof similar to Theorem 7 could be applied to the number of nodes in $BZRG_p$. However, the tree associated with $BZRG_p$ grows by a complete level with probability p, instead of one node at each time. Thus, the current depth of the tree has to be known by the function in order to add the corresponding number of nodes, hence $N = pt(p, 0)$ and $t(p, k) = pt(p, k + 1) * 2^k$.

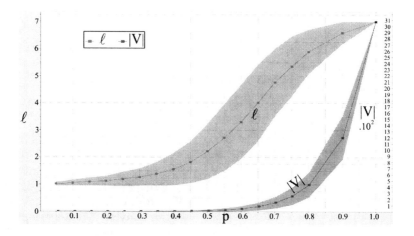

Fig. 7. The average path length has a slow growth compared to the number of nodes, as in $BZRG_p$. We show the values averaged accross simulations and the standard deviation. *Color online.*

In the previous model, we showed that the expected average distance had a constant growth whereas the expected number of nodes had an exponential growth. Thus, the gap between the two could be arbitrarily large. Using simulations reported in Figure 7, we show that the same effect holds in this model.

6   Conclusion and Future Work

We proved a close-form formula for the average distance in ZRG_t. We proposed two stochastic versions, and showed that they had a first-order phase transition at the same critical probability. In the recent years, we have witnessed many complex network models in which nodes are added at each time step. The graph-theoretic and probabilistic techniques illustrated in our paper can thus be used to rigorously prove the behaviour of models.

References

1. Schnettler, S.: Social Networks 31, 165 (2009) ISSN 03788733
2. Giabbanelli, P., Peters, J.: Technique et Science Informatiques (2010) (to appear)
3. Watts, D.J.: Small worlds: the dynamics of networks between order and randomness. Princeton University Press, Princeton (1999)
4. Davidsen, J., Ebel, H., Bornholdt, S.: Phys. Rev. Lett. 88 (2002)
5. Zhang, Z., Chen, L., Zhou, S., Fang, L., Guan, J., Zou, T.: Phys. Rev. E 77 (2008)
6. Zhang, Z., Rong, L., Guo, C.: Physica A 363, 567 (2006)
7. Giabbanelli, P., Mazauric, D., Perennes, S.: Proc. of the 12th AlgoTel (2010)
8. Zhang, Z., Rong, L., Comellas, F.: J. of Physics A 39, 3253 (2006)
9. Cohen, R., Havlin, S.: Phys. Rev. Lett. 90 (2003)

Traffic Congestion on Clustered Random Complex Networks

Thiago Henrique Cupertino and Liang Zhao

Institute of Mathematical Sciences and Computing, Av. Trabalhador São-carlense
400, São Carlos, SP, Brasil
{thiagohc,zhao}@icmc.usp.br

Abstract. In this work we study the traffic-flow on clustered random complex networks. First, we derive a mathematical model to determine the congestion phase-transition point. This point is defined as the abrupt transition from a free-flow to a congested state. Second, we study the influences of different cluster sizes on the traffic-flow. Our results suggest that the traffic of centralized cluster network (a network which has a big central cluster surrounded by clusters with significantly smaller sizes) is less congesting than balanced cluster network (a network with clusters of approximately the same size). These results may have practical importance in urbanization planning. For example, using the results of this study, the increasing of satellite cities' sizes surrounding a big city should be well controlled to avoid heavy traffic congestion.

1 Introduction

A rapidly increase on the interest studying networks with complex structures have been observed since about 12 years ago. The works vary on the areas of dynamical processes, analytical properties, structural models and so on [1,2,3]. Examples of such complex networks on the real world are the Internet, the WWW, P2P networks, social networks, telecommunication systems, biological networks (e.g. neural networks, protein-protein interaction networks and neural networks), traffic networks, paper author citations networks etc [4,5].

On some kind of networks such as the Internet, transportation networks and power grids, the free traffic-flow of the characteristic element (e.g., data, goods and signals, respectively) are crucial for the well-functioning of the whole system. Furthermore, are also crucial for the society as we depend more and more on these systems. Motivated by that, several studies have been done considering traffic-flow in complex networks [6,7,8,9,10,11,12,13]. Many of the previous works on this area consider networks with regular or homogeneous structures. However, some works revealed that many real networks possesses different kinds of topologies like scale-free, small-world and communities [1,2]. Moreover, a previous work [9] has shown that for two networks with the same average connectivity, node processing capacity, and total number of nodes, if their topological structures are different, the traffic phase-transition point can be significantly different, indicating that traffic congestion depends sensitively on network structure. On the

L. da F. Costa et al. (Eds.): CompleNet 2010, CCIS 116, pp. 13–21, 2011.

same direction, in this work we study the traffic-flow and congestion on clustered complex networks.

As a specific structure of complex networks, the clustered complex networks or the community structured complex networks are characterized by groups of nodes densely connected inside the group while the connections between different groups are sparse [14]. The structures of communities and between communities are important for the demographic identification of network components and the function of dynamical processes that operate on these networks (e. g., the spread of opinions and deceases) [15,16].

Another important feature of many complex networks is that its properties can change over time. In a period of time, new nodes or links can be added or redistributed on a given network. As examples of these evolving networks are the social networks, where opinion formation is mediated by social interactions, consequently taking place on a network of relationships and at the same time influencing the structure of the network and its evolution [17]. Another example could be a cellular network composed of wireless mobile devices, where a group of nodes is defined by their geographical positions (devices next to each other are connected by the same tower/antenna), and the users are in a continuous movement which changes the network structure [18].

In [13], it has been shown that the better defined the communities are, the more affected packet transport becomes. Now, we are interested in studying the influence of communities sizes in the packet transportation. By combining the clustering and the evolving features of complex networks we are able to study some properties of evolving clustered complex networks. Specifically, in this work we are also interested on the traffic-flow on networks where the cluster size changes over time. From the results of our simulations we verified that when a network have clusters of similar sizes the traffic is more congested than when there are a central cluster and peripheral smaller ones. An analogy with these results can be made using the vehicle traffic between a big city and the satellite cities around it. Our study suggests that the traffic flow can be less congested if a significant difference in size between the big city and its satellite cities can be maintained, in other words, the increasing of satellite cities should be well controlled in urbanization process.

The remainder of the paper is organized as follows. In Sec. 2, we describe the traffic-flow model used in this work. In Sec. 3, we carry out an analysis of the clusterized traffic model to determine the congestion phase-transition point. In Sec. 4, we give numerical support for our analysis and study the influence of different cluster sizes on the traffic-flow. Finally, in Sec. 5 we give some brief conclusions and comments.

2   Traffic-Flow Model

As the model described in [9,10], the study of traffic-flow on clustered complex networks in this paper considers two parameters: the information generation rate λ and a control parameter β, that measures the capacity of nodes to process information. The capacity of each node for processing and sending information are proportional to the number of links connected to the node (node's degree). The quantity of interest is the critical value for λ at which a phase transition occurs from free to congested traffic-flow. In particular, for $\lambda \leq \lambda_c$, the numbers of created and delivered packets are balanced, resulting in a steady state, or free traffic-flow. For $\lambda > \lambda_c$, congestions occur in the sense that the number of accumulated packets increases in time, due to the fact that the capacities of nodes for delivering packets are limited. We are interested in determining the phase-transition point, given a clustered complex network, in order to address which kind of clustering is more susceptible to phase transition and therefore traffic congestion.

We use the same traffic-flow model described in [9,10], which is based on the routing algorithm in computer networks. The capacities for processing information are different for different nodes, depending on the numbers of links passing through them. The routing algorithm consists on the following steps:

(1) At each time step, the probability for node i to generate a packet is λ.
(2) At each time step, a node i delivers at most C_i packets one step toward their destinations, where $C_i = (1 + int[\beta k_i])$ and $0 < \beta < 1$ is a control parameter, k_i is the degree of node i and B_i is its betweenness (calculated using Newman's algorithm [19]).
(3) A new created packet is placed at the end of the queue of its creating node. At the same time, another node on the network is chosen as a destination node and the packet is forwarded, on next time steps, along the shortest path between these two nodes. If there are several shortest paths for one packet, it is chosen the one having the shortest queue length.
(4) A packet, once reaching its destination node, is removed from the traffic.

Although traffic-flow rules defined in this model are simple, they capture the essential characteristics of traffic-flow in real networks. Our aim is to determine the relationship between the network community structure and the critical packet-generating rate λ_c, which is described in the next Section.

3 Theoretical Analysis

In our work, we use the following random clustered complex network model: N nodes are divided into M groups, where each group has $n = N/M$ nodes. Inside the same group, a pair of nodes is connected randomly with probability ρ_{in}, and nodes belonging to different groups are connected randomly with probability ρ_{out}. For a clustered complex network, the number of interconnections is typically far less than the number of intra-connections. So, we study the cases in which $\rho_{out} \ll \rho_{in}$, basically $\rho_{out} \approx \rho_{in}/10$.

As stated in [9], the critical point for the packet generation rate on a complex network is defined by:

$$\lambda_c = \frac{(1 + int[\beta k_{max}])(N - 1)}{B_{max}}, \qquad (1)$$

where k_{max} and B_{max} are, respectively, the degree and the betweenness of the node with the largest betweenness on the network.

The problem here is to find these measures analytically for a given network structure. In our case, we can promptly find approximated measures for randomly clustered complex networks. To easily understand, consider a clustered network with N nodes and 2 communities, each one having $n = N/2$ nodes. Given that $\rho_{out} \ll \rho_{in}$, there will be just a few nodes connecting both communities and, consequently, these few nodes will be loaded with all the traffic between the communities. So, considering that $n \gg n\rho_{out}$, i.e., the cluster size is much greater than the averaged number of nodes with interconnections, we can consider that these nodes will have the largest betweennesses of the network and so they will be the bottleneck for the traffic-flow.

To find B_{max}, we divide the analysis in two parts. The first concerns the betweenness inter-clusters and the second, intra-clusters. Considering that there are just a few nodes with interconnections, the averaged betweenness for them is found by dividing the number of shortest paths inter-clusters by the averaged number of that kind of nodes, i.e.:

$$B_{inter} = \frac{n(N-n)}{n\rho_{out}} = \frac{N-n}{\rho_{out}} \tag{2}$$

As we are studying random clusters, the intra-cluster betweenness can be found with good approximation by averaging the total betweenness inside a cluster, i.e.:

$$B_{intra} = \frac{Dn(n-1)}{n} = D(n-1), \tag{3}$$

where D is the averaged shortest path inside the clusters. This approximation is reasonable because each bottleneck node is randomly selected and thus it has high probability that it degree is near the average degree of the cluster. Consequently, its betweenness should be close to the average betweenness of the cluster.

So, the maximum betweenness is given by:

$$B_{max} = B_{inter} + B_{intra} = \frac{N-n}{\rho_{out}} + D(n-1), \tag{4}$$

and the critical packet generating rate (Eq. 1) becomes:

$$\lambda_c = \frac{(1 + int[\beta < k >])(N-1)}{\frac{N-n}{\rho_{out}} + D(n-1)}, \tag{5}$$

where we used the averaged degree $< k >$ of the network.

Now, consider a clustered random network where initially all clusters have the same size n. After a while, one cluster evolves by expanding its size n while the other clusters become smaller. This process can be understood by creating another network with the same parameters, but with different communities' sizes. So, applying Eq. 5 to the biggest cluster, it can be noted that λ_c depends on

n through two measures: B_{inter} $((N-n)/(\rho_{out}))$ and B_{intra} $(D(n-1))$. As n increases, B_{inter} decreases linearly (and so λ_c increases and traffic becomes less congested) until the limit $n = N$ when it vanishes. On the other hand, as n increases B_{intra} increases logarithmically ($D \sim ln(n)$ [21]) (and so λ_c decreases and traffic becomes more congested). Also, from the network structure we have $B_{inter} \gg B_{intra}$ and, therefore, from these two opposite consequences we can expect more influence from B_{inter} on increasing λ_c (attenuating traffic congestion) when a cluster is expanding.

4   Numerical Support

In order to characterize the transition from free traffic-flow to congestion, we use the order parameter $\eta = lim_{t \longrightarrow \infty} < \Delta\Theta > /\lambda\Delta t$ [20] where $< \Delta\Theta >= \Theta(t + \Delta t) - \Theta(t)$, $\Theta(t)$ is the total number of packets in the network at time t, and $< ... >$ indicates average over a time window of length Δt. For $\lambda < \lambda_c$ the network is in a free flow state. In this case, we have $< \Delta\Theta > \approx 0$ and $\eta \approx 0$. For $\lambda > \lambda_c$, traffic congestion occurs so that $\Delta\Theta$ increases with Δt.

Figure 1 shows λ versus η for various clustered complex networks. For networks with $N = 1000$ nodes with clusters of $n = 200$ nodes, various probability connection rates are studied. In these results, we simulated the probability of interconnections ρ_{out} ranging from 0.01 to 0.13, while the probabilities for intra-connections are $\rho_{in} = 1 - \rho_{out}$. As expected from Eq. 5, the network is more congested when the connections between clusters are sparse. As ρ_{out} increases, the clusters of the network become more connected and the traffic is less congested.

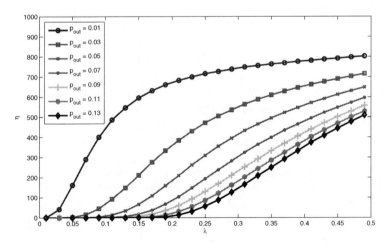

Fig. 1. Order parameter η vs the packet-generation rate λ for five-cluster networks of different connections probabilities. For each data point the network size is $N = 1000$, $n = 200$, $< k >= 8$ and $\beta = 0.1$. 30 different scenarios were averaged.

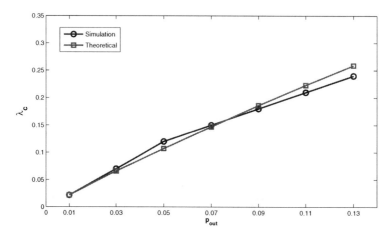

Fig. 2. Comparison of the critical generating rate λ_c from theoretical prediction (squares) and from numerical simulations (circles). For each data point the network size is $N = 1000$, $n = 200$, $< k >= 8$ and $\beta = 0.1$. 30 different scenarios were averaged.

Figure 2 shows the comparison of the critical generating rate from theoretical prediction (calculted by using Eq. 5) and from numerical simulations. These simulations are the same as the Fig. 1 and so we use the same parameters. One can note here that the theoretical model given by Eq. 5 can get a good estimation of the critical packet generating rate for the clustered complex networks under study.

Two different simulations are performed to study the influence of the changing of the network cluster structure on the traffic-flow. The first simulation considers a network of size $N = 400$ with two clusters. Initially, both clusters have the same number of nodes $n = 200$. As the time passes, one cluster starts growing while the other one becomes smaller, until the limit of $n_1 = 360$ and $n_2 = 40$ nodes, respectively. This situation is shown by Fig. 3: when a network maintains the same number of nodes while its communities are changing over time, the phase-transition point from the free flow state to the congested state changes. In these cases, the most congested network is the one which has the most similar cluster sizes. The second simulation considers an environment where both clusters are expanding over time, but one increases quicker than the other. Here, the network starts with two clusters: one with $n_1 = 200$ nodes and the other with $n_2 = 40$ nodes. At each iteration, the former expands with a rate $\alpha_1 = 5\%$ while the later, with $\alpha_2 = 20\%$. At the last iteration, the maximum sizes for the first and the second cluster are $n_1 = 353$ and $n_2 = 357$, respectively. As shown by Fig. 4, two conclusions can be made. First, the values of η naturally increase through the iterations as the total network size N increases and more packets are being created. But here the second observation is more important: as the previous simulation, when the clusters have similar sizes the network is more congested. So, from the two scenarios simulated, the first with fixed N and the

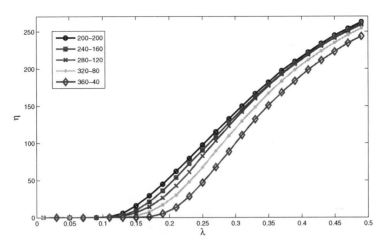

Fig. 3. Order parameter η vs the packet-generation rate λ for two-cluster networks with different cluster sizes. Initially, both clusters have the same number of nodes $n = 200$. On the last scenario one cluster has $n = 360$ nodes while the other one has only $n = 40$ nodes. The connection probabilities are $p_{out} = 0.05$ and $p_{in} = 0.95$. For each data point the network size is $N = 400$, $< k >= 8$ and $\beta = 0.1$. 30 different scenarios were averaged.

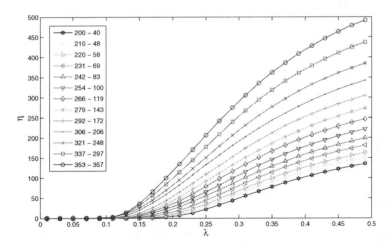

Fig. 4. Order parameter η vs the packet-generation rate λ for two-cluster networks with different cluster sizes. Initially, one cluster has $n_1 = 200$ and the other, $n_2 = 40$ nodes. The first cluster increases with rate $\alpha_1 = 5\%$ while the other, with $\alpha_2 = 20\%$. The connection probabilities are $p_{out} = 0.05$ and $p_{in} = 0.95$. For each data point the average degree is $< k >= 8$ and $\beta = 0.1$. 30 different scenarios were averaged.

second with an increasing N, we observed the same influence of the cluster sizes on the network traffic-flow.

5   Conclusions

In this work, we studied the traffic-flow in complex networks. Motivated by practical applications and previous works [9,10], our primary interest was to investigate the congestion process on a clustered structure. We first derived a mathematical analysis to determine the phase-transition congestion point and then we shown numerical results. The simulation results suggest that when a network has clusters with different sizes, e. g., a central cluster surrounded by clusters with significantly smaller sizes, the traffic is less congested than when the clusters have approximately the same size. To our knowledge, this is the first time a study on how sub-network sizes influence the traffic-flow in complex networks. We believe that these results have practical importance for controling flow and designing networks in which there is some kind of traffic and that present a clustered structure. For example, in order to decrease traffic congestion in a metropolitan region of a big city, the satellite cities or communities should not increase rapidly, in other words, it should have more small satellite cities rather than few big satellite cities. As future works, it should be addressed more complex network structures, i.e., having a more number of clusters and different topologies as scale-free and small-world.

Acknowledgments. T.H.C. is supported by FAPESP under Grant No. 2009/02036-7. L.Z. is supported by CNPq under Grant No. 302783/2008-3.

References

1. Barabási, A.-L., Albert, R.: Science 286, 509 (1999)
2. Albert, R., Barabási, A.-L.: Rev. Mod. Phys. 74, 47 (2002)
3. Newman, M.E.J.: SIAM 45, 167 (2004)
4. Strogatz, S.H.: Nature 410, 268 (2001)
5. Dorogovtsev, S.N., Mendes, J.F.F.: Evolution of Networks: From Biological Nets to the Internet and WWW. Oxford University Press, Oxford (2003)
6. Guimerà, R., Díaz-Guilera, A., Vega-Redondo, F., Cabrales, A., Arenas, A.: Phys Rev. Lett. 89, 248701 (2002)
7. Woolf, M., Arrowsmith, D.K., Mondragón, R.J., Pitts, J.M.: Phys. Rev. E 66, 046106 (2002)
8. Valverde, S., Solé, R.V.: Physica A 312, 636 (2002)
9. Zhao, L., Lai, Y.-C., Park, K., Ye, N.: Phys. Rev. E 71, 026125 (2005)
10. Zhao, L., Cupertino, T.H., Park, K., Lai, Y.-C., Jin, X.: Chaos 17, 043103 (2007)
11. Danila, G., Sun, Y., Bassler, K.E.: Phys. Rev. E 80, 066116 (2009)
12. Yang, R., Wang, W.-X., Lai, Y.-C., Chen, G.: Phys. Rev. E 79, 026112 (2009)
13. Danon, L., Arenas, A., Díaz-Guilera, A.: Phys. Rev. E 77, 036103 (2008)
14. Girvan, M., Newman, E.J.: Proc. Natl. Acad. Sci. 99, 7821 (2002)
15. Porter, M.A., Onnela, J.-P., Mucha, P.J.: Notices of the American Mathematical Society 56, 1082 (2009)

16. Fortunato, S.: Physics Reports 486, 75 (2010)
17. Iñiguez, G., Kertész, J., Kaski, K.K., Barrio, R.A.: Phys. Rev. E 80, 066119 (2009)
18. Borgnat, P., Fleury, E., Guillaume, J.-L., Robardet, C., Scherrer, A.: Proceedings of NATO ASI 'Mining Massive Data Sets for Security'. In: Fogelman-Soulié, F., Perrotta, D., Piskorski, J., Steinberg, R. (eds.) NATO Science for Peace and Security Series D: Information and Communication Security, vol. 42, p. 198. IOS Press, Amsterdam (2008)
19. Newman, M.E.J.: Phys. Rev. E 64, 016132 (2001)
20. Arenas, A., Díaz-Guilera, A., Guimerà, R.: Phys Rev. Lett. 86, 3196 (2001)
21. Newman, M.E.J., Strogatz, S.H., Watts, D.J.: Phys. Rev. E 64, 026118 (2001)

Fully Generalized Graph Cores

Alexandre P. Francisco and Arlindo L. Oliveira

INESC-ID / CSE Dept, IST, Tech Univ of Lisbon
Rua Alves Redol 9, 1000-029 Lisboa, PT
{aplf,aml}@inesc-id.pt

Abstract. A core in a graph is usually taken as a set of highly connected vertices. Although general, this definition is intuitive and useful
for studying the structure of many real networks. Nevertheless, depending on the problem, different formulations of graph core may be required,
leading us to the known concept of generalized core. In this paper we
study and further extend the notion of generalized core. Given a graph,
we propose a definition of graph core based on a subset of its subgraphs
and on a subgraph property function. Our approach generalizes several
notions of graph core proposed independently in the literature, introducing a general and theoretical sound framework for the study of fully
generalized graph cores. Moreover, we discuss emerging applications of
graph cores, such as improved graph clustering methods and complex
network motif detection.

1   Introduction

The notion of k-core was proposed first by Seidman [1] for unweighted graphs
in 1983. We say that a subgraph H of a graph G is a k-core or a *core of order*
k if and only if H is a maximal subgraph such that $d(v) \geq k$, for all vertices
v in H and where $d(v)$ is the degree of v with respect to H. Although general,
this definition is intuitive and useful for the study of the structure of many real
networks, with the first applications appearing in the area of social sciences [2].
More recently Batagelj and Zaveršnik [3] introduced the notion of generalized
cores, allowing the definition of cores based on a vertex property function. In
this paper, we further extend the concept of generalized core. Instead of vertex
property functions, we consider subgraph property functions in general, leading
to fully generalized cores. Moreover, we show that many notions of graph core
and densely connected subgraphs proposed independently in the literature can
be defined as fully generalized cores.

Given a graph $G = (V, E)$, possibly weighted, the problem consists of finding
highly connected subgraphs. This problem is well known in graph clustering, where
these subgraphs play the role of cluster or community cores. In particular, given a
core for G, we can obtain a clustering for G by taking each core connected component as a seed set and by applying local partitioning methods [4,5]. The notion of
core has also been used in the context of multilevel schemata for graph clustering,
where coarsening schemata were found to be closely related to the problem of core

L. da F. Costa et al. (Eds.): CompleNet 2010, CCIS 116, pp. 22–34, 2011.

enumeration [6,7]. The main idea behind most of the existing notions is to merge the vertices that are more similar, namely in what concerns connectivity. Since we can define several vertex similarity scores and we can take different merging strategies, there are many possible definitions of core. The notion of fully generalized core proposed in this paper becomes particularly useful in this context.

Cliques and, in particular, the clique percolation method by Palla *et al.* [8] to detect densely connected, and possibly overlapping, clusters or communities on networks, are also related to graph cores. A clique is a complete graph and, if it has k vertices, then it is called a k-clique. The idea behind the clique percolation is that the edges within a cluster or community are likely to form cliques, *i.e.*, highly connected subgraphs. Conversely, the edges among clusters should not form cliques. In this context, we say also that two k-cliques are adjacent if they share $k - 1$ vertices and, thus, a k-clique community is the largest connected subgraph obtained by the union of adjacent k-cliques. The method can be extended to weighted graphs either by considering a threshold on edge weights or a threshold on clique weights, defined for instance as the geometric mean of the weights of all edges [9]. Another maximal clique based approach was recently proposed by Shen *et al.* [10] to uncover both the overlapping and hierarchical community structures of networks. This method uses an agglomerative approach, merging pairs of maximal cliques that maximize a given similarity score. The main drawbacks of these methods are that detecting maximal cliques is an NP-hard problem, even though the authors found that the method is fast in practice for sparse networks, and that taking cliques as cluster building blocks may be an assumption too strong for many real networks. As pointed out by Saito *et al.* [11], methods based on the computation of graph cores or its extensions, *e.g.* k-dense clusters, can be better than methods based on the computation of maximal cliques. In particular, both k-cores and k-dense cores are less restrictive than k-cliques. Here, we analyse also k-cliques, k-clique percolations and k-dense clusters since these notions are particular cases of fully generalized cores.

The notion of fully generalized core introduced in this paper is also closely related with network motifs, allowing for composed network motifs. In fact, we can think of fully generalized cores as subgraphs formed by merging together highly connected motifs. The role of subgraph property functions is precisely to evaluate motif connectedness with respect to some criteria. Recent works in network analysis have made it clear that large complex natural networks reveal many local topological patterns, regarded as simple building blocks in networks, and named motifs. These characteristic patterns have been shown to occur much more frequently in many real networks than in randomized networks with the same degree sequence. For example, Milo *et al.* [12] discovered frequent characteristic local patterns on biological networks, *i.e.*, network motifs, observing that certain motifs are more common on biological networks than in other complex networks, revealing basic structural elements of such networks. Many efforts were done in order to understand the importance of network motifs [13,14] and promising results were achieved, in spite of the rather limited network motifs that were used. For instance, Saito *et al.* [13] used only five predefined network motifs of size three and Albert *et al.* [14] used only four predefined small network motifs. Note that many relevant processes in biological networks correspond to the mesoscale and, therefore, it will be interesting to study larger network motifs. Most of current network motif finding algorithms [12,15] are enumeration based and limited to the extraction of smaller network motifs. The first reason is that the number of potential network motifs increases exponentially with the motif size [16]. A second one is that interesting motifs occur repeatedly in a given network but not in other networks, namely in randomized ones [12]. A third reason is that finding a given motif is closely related to the subgraph isomorphism problem. These reasons make the application of enumeration based algorithms unpractical when we consider mesoscale network motifs. Although different from motifs, we may want to study the occurrence of graphlets instead. Usually, graphlets must be induced subgraphs while motifs may be partial subgraphs. See for instance the recent work by Milenković *et al.* [17]. Graphlet frequency and degree distribution has been shown to provide good network signatures, becoming useful for the study of complex networks and for comparing them against proposed network models. As mentioned for motifs, the notion of fully generalized core is also useful to study graphlet composed cores.

2 Preliminaries

A *graph* or an *undirected graph* G is a pair (V, E) of sets such that $E \subseteq V \times V$ is a set of *unordered* pairs. The elements of V are the *vertices* and the elements of E are the *edges*. In this paper we assume that a graph does not have several edges between the same two vertices, *i.e.*, it does not have multiple edges, or edges that start and end at same vertex, *i.e.*, loops. When E is a set of *ordered* pairs we say that G is a *directed graph*. In this case the edge (u, v) is different from the edge (v, u) since they have different directions.

Given a graph $G = (V, E)$, the vertex set of G is denoted by $V(G)$ and the edge set of G denoted by $E(G)$. Clearly $V(G) = V$ and $E(G) = E$. The number of vertices of G is its *order*, denoted either by $|V|$ or by $|G|$, and its number of edges is denoted by $|E|$. We say that a graph G is *sparse* if $|E| \ll |V|^2$. Two vertices $u, v \in V(G)$ are *adjacent* or *neighbors* if (u, v) is an edge, *i.e.*, $(u, v) \in E(G)$. Given a vertex $v \in V$, its set of neighbors is denoted by $N_G(v)$, or by $N(v)$ when G is clear from the context. The number of neighbors of v is its *degree* denoted by $d_G(v)$, or by $d(v)$ or d_v when G is clear from the context, *i.e.*, $d(v) = |N(v)|$. Given $V' \subseteq V(G)$, $d(V')$ denotes the sum of $d(v)$ for each $v \in V'$, *i.e.*, $d(V') = \sum_{v \in V'} d(v)$.

Let us now recall some graph properties. A graph G is *complete* or a *clique* if all vertices of G are pairwise adjacent. Usually, if G is complete and $|G| = n$, we denote G by K_n. Two graphs G and G' are *isomorphic*, denoted by $G \simeq G'$, if there is a bijection $\eta : V(G) \to V(G')$ such that $(u, v) \in E(G)$ if and only if $(\eta(u), \eta(v)) \in E(G')$, for all $u, v \in V$. Sometimes we are only interested in the notion of subgraph. G' is said to be a *subgraph* of G and G a *supergraph* of G' if $V(G') \subseteq V(G)$ and $E(G') \subseteq \{(u, v) \in E(G) \mid u, v \in V(G')\}$. G' is said to be a *proper subgraph* if $V(G') \subsetneq V(G)$. Given $V' \subseteq V(G)$, the subgraph *induced* by V' is the graph $G' = (V', E')$ where $E' = \{(u, v) \in E(G) \mid u, v \in V'\}$.

A *weighted graph* G is a tuple (V, E, w) where V and E form a graph $G = (V, E)$ and $w : E \to \mathbb{R}$ is a function that assigns to each edge $e \in E$ a weight $w(e)$. Note that we could also assign weights to the vertices or even arbitrary labels to both vertices and labels.

A *vertex similarity function* σ maps each pair of vertices to a positive real value, $\sigma : V^2 \longrightarrow \mathbb{R}_0^+$. Note that σ may be different from, although usually related to, the edge weight function w. Since σ reflects the similarity between two vertices $u, v \in V$, we usually say that u and v are the more similar the higher the value $\sigma(u, v)$. Moreover, we ignore pairs of vertices $u, v \in V$ for which $\sigma(u, v)$ is 0.0. The choice of the σ functions will always depend on the problem under study. For instance, we can simply use the vertex degree or the edge weights, if a suitable edge weight function w is provided. But, in general, these are not enough. For instance, we can consider a structural similarity function based on the cosine similarity. Note that we could start with other similarity functions, *e.g.*, with the Jaccard-Tanimoto index [18,19]. Let w be the edge weight function. Given two connected vertices $(u, v) \in E$, their *structural similarity* $\sigma(u, v)$ is given by

$$\sigma(u, v) = \frac{2\, w(u, v) + \sum_{x \in N(u) \cap N(v)} w(u, x)w(v, x)}{\sqrt{1 + \sum_{x \in N(u)} w(u, x)^2} \sqrt{1 + \sum_{x \in N(v)} w(v, x)^2}}. \tag{1}$$

This equation reflects the cosine similarity between the neighborhoods of u and v. The term $2\, w(u, v)$ in the numerator and the 1's in the denominator were introduced to reflect the connection between u and v, being the only difference with respect to the usual definition of cosine similarity. In particular, if we extend this definition to all distinct pairs of vertices $u, v \in V$ or if we consider directed graphs, we may want to drop these terms. The version of Eq. (1) for unweighted graphs was first proposed by Xu *et al.* [20]. The similarity function σ as defined in Eq. (1) takes values in $[0, 1]$ and, given $(u, v) \in E$, $\sigma(u, v)$ grows as u and v share more neighbors. If u and v share all neighbors with equal weights, $\sigma(u, v)$ is 1.0. In particular, $\sigma(u, v)$ is 1.0 even if u and v share all neighbors through equal lowly weighted edges. In order to distinguish common neighbors connected through lowly weighted edges from common neighbors connected through highly weighted edges, we can compute the average weight among the common neighbors

$$\overline{w}(u, v) = \frac{w(u, v) + \sum_{x \in N(u) \cap N(v)} w(u, x) + w(v, x)}{1 + |N(u) \cap N(v)|} \tag{2}$$

and redefine σ as the product of $\overline{w}(u, v)$ by Eq. (1). Note that we may consider other terms instead of the weight average. For instance we could compute the maximum weight. Note also that σ as redefined above only takes values in $[0, 1]$ if w also takes values in $[0, 1]$.

We say that the subgraph H of G induced by $C \subseteq V$ is a *k-core* or a *core of order* k if and only if $d_H(v) \geq k$, for all $v \in C$, and H is a maximal subgraph with this property. The notion of k-core was proposed first by Seidman [1] for unweighted graphs. Usually, by abuse of nomenclature, each connected component of H is also called a k-core. More recently, Batagelj and Zaveršnik [3] proposed a generalization of the notion of core, allowing the use of other properties of vertices than their degree. A *vertex property function* p on V is such that $p : V \times 2^V \longrightarrow \mathbb{R}$. Given $\tau \in \mathbb{R}$, a subgraph H of G induced by $C \subseteq V$ is a τ-*core with respect to* p, or a p *core at level* τ, if $p(v, C) \geq \tau$ for all $v \in C$ and H is a maximal subgraph with this property.

Given a vertex property function p, we say that p is *monotone* if and only if, given $C_1 \subseteq C_2 \subseteq V$, $p(v, C_1) \leq p(v, C_2)$ for all $v \in V$. Then, given a graph G, a monotone vertex property function p and $\tau \in \mathbb{R}$, we can compute the τ-core of G with respect to p by successively deleting vertices with p value lower than τ:

1. set $C \leftarrow V$;
2. while exists $v \in C$ such that $p(v, C) < \tau$, set $C \leftarrow C \setminus \{v\}$.

Theorem 1. *Given a graph G, a monotone vertex property function p and $\tau \in \mathbb{R}$, the above procedure determines the τ-core with respect to p.*

Corollary 1. *Given a monotonic vertex property function p and $\tau_1, \tau_2 \in \mathbb{R}$ such that $\tau_1 < \tau_2$, the cores are nested, i.e., $C_2 \subseteq C_1$.*

These two results are due to Batagelj and Zaveršnik [3] and are particular cases of the more general results presented in this paper.

We can devise many vertex property functions. Here, we discuss three examples. Given a graph $G = (V, E)$, we can recover the classical definition of k-core by defining the vertex property function

$$p(v, C) = d_H(v), \tag{3}$$

where H is the subgraph of G induced by C. Thus, given $k \in \mathbb{N}$, a k-core with respect to p is precisely a classical k-core as defined by Seidman. Given a vertex similarity function σ, we can extend Eq. (3) as

$$p(v, C) = \sum_{u \in N(v) \cap C} \sigma(v, u). \tag{4}$$

Note that, taking σ as the weight function w, Eq. (4) is the natural extension of the k-core notion to weighted graphs leading to the notion of τ-core with $\tau \in \mathbb{R}$.

As in Eq. (1), the similarity function may already evaluate how strongly a vertex is connected to its neighbors. Thus, we may prefer the property function

$$p(v, C) = \max_{u \in N(v) \cap C} \sigma(v, u). \tag{5}$$

In this case, for $\tau \in \mathbb{R}$, all vertices v in the τ-core H are connected to some other vertex u in H such that $\sigma(u, v) \geq \tau$. With the vertex property function (5) the problem of finding cores becomes closely related to graphic matroids [21,22]. In particular, taking σ as the weight function, a τ-core H is the maximal subgraph of G such that all edges in a maximum spanning forest of H have weight higher than τ. There are two efficient and well known approaches to enumerate the cores. We can sort the pairs of distinct vertices $u, v \in V$ by decreasing order of

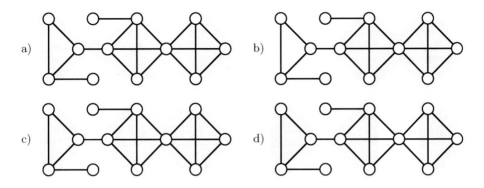

Fig. 1. Cores for different vertex property functions: a) 2-core with respect to the vertex property function (3); b) 3-core with respect to Eq. (3); d) 0.75-core with respect to Eq. (5); d) 0.85-core with respect to Eq. (5)

$\sigma(u, v)$ and iteratively merge them to form cores, which is the principle behind the algorithm of Kruskal [23]. Or we can iteratively visit each vertex u and merge it with the neighbor v that maximizes $\sigma(u, v)$, an approach related to the algorithm of Borůvka [24]. Thus, as is well known, both these approaches take $O(m \log n)$ where n is the number of vertices and m is the number of pairs such that $\sigma(u, v) > 0$. Note also that, if we consider the algorithm of Kruskal, we can get the full core hierarchy in a single run. We just need to store the cores for the thresholds we are interested in or, if preferred, the full dendrogram.

The examples of vertex property functions above are all monotonic, *i.e.*, $p(v, C_1) \leq p(v, C_2)$ for $C_1 \subseteq C_2 \subseteq V$. This is a straightforward consequence from the fact that, for the first two examples, σ is always positive and p is additive with respect to C and, in the third example, that the maximum can only increase as C grows.

3   Fully Generalized Cores

Let us now extend the notion of generalized core. Recently, Saito *et al.* [11] studied k-dense communities, where each pair of adjacent vertices must share at least $k-2$ neighbors. This is clearly related to the k-core notion. The difference is that they consider pairs of connected vertices instead of a single vertex. Moreover, Saito *et al.* pointed out that an extension would be the use of cliques, in general, instead of vertices or edges. Here, we further exploit these ideas and we propose an extension of generalized cores, allowing the evaluation of density for any subgraph. Let $G = (V, E)$ be a graph and let 2^G denote the set of subgraphs of G. Given $\mathcal{M} \subseteq 2^G$ a set of subgraphs of G, for instance a set of motifs, a *subgraph property function* p over $\mathcal{M}$ is such that $p : \mathcal{M} \times 2^G \longrightarrow \mathbb{R}$. We say that p is *monotone* if and only if the following conditions hold:

1. if H_1 is subgraph of $H_2 \in 2^G$, then $p(M, H_1) \leq p(M, H_2)$, for all $M \in \mathcal{M}$;
2. if $L_1 \in \mathcal{M}$ is subgraph of $L_2 \in \mathcal{M}$, then $p(L_1, H) \geq p(L_2, H)$, for all $H \in 2^G$.

The first condition is the generalization of the monotonicity condition discussed in the previous section. The second condition will allow us to refine cores with respect to p by changing the set of subgraphs $\mathcal{M}$, as stated in Proposition 1 and depicted in Fig. 2.

Let H be a subgraph of G, *i.e.*, $H \in 2^G$. We define $\mathcal{M}(H)$ as the set of subgraphs of H in $\mathcal{M}$, *i.e.*, $\mathcal{M}(H) = \mathcal{M} \cap 2^H$. Given $\tau \in \mathbb{R}$, H is a τ-*core with respect to* p, or a p *core at level* τ, if

1. $V(H) \subseteq \bigcup_{M \in \mathcal{M}(H)} V(M)$,
2. $p(M, H) \geq \tau$, for all $M \in \mathcal{M}(H)$,
3. and H is a maximal subgraph of G with properties 1 and 2.

The first condition states that H must be a subgraph of G induced by a set of subgraphs in $\mathcal{M}$. The second condition ensures that all subgraphs of H in $\mathcal{M}$ are densely connected within H and with respect to p. Finally, the third condition requires that H is maximal, *i.e.*, that there is not any τ-core H' with respect to p such that H is subgraph of H'. As before, by abuse of nomenclature, each connected component of H may also be called a core.

Given a graph G, $\mathcal{M} \subseteq 2^G$, a monotonic subgraph property function p over $\mathcal{M}$ and $\tau \in \mathbb{R}$, we can compute the τ-core H of G with respect to p as follows:

1. set H as the subgraph of G induced by $\bigcup_{M \in \mathcal{M}} V(M)$, *i.e.*, initialize H as the subgraph of G induced by the vertices of all subgraphs in $\mathcal{M}$;
2. while exists $M \in \mathcal{M}(H)$ such that $p(M, H) < \tau$, set H as the subgraph of G induced by $\bigcup_{M' \in \mathcal{M} \setminus \{M\}} V(M')$, *i.e.*, remove M from the list of subgraphs under consideration.

Theorem 2. *Given a graph G, $\mathcal{M} \subseteq 2^G$, a monotonic subgraph property function p over $\mathcal{M}$ and $\tau \in \mathbb{R}$, the above procedure determines the τ-core wrt p.*

Proof. Let H be the core returned by the procedure. We must show that

1. $p(M, H) \geq \tau$, for all $M \in \mathcal{M}(H)$;
2. H is maximal and independent of the order of deletions, *i.e.*, unique.

It is clear that 1 holds since all subgraphs M such that $p(M, H) < \tau$ are deleted in the procedure. Let us show that 2 also holds by absurd. Suppose that exists H' also determined by the above procedure, but such that $H' \neq H$. Thus, we have either $\mathcal{M}(H') \setminus \mathcal{M}(H) \neq \emptyset$ or $\mathcal{M}(H) \setminus \mathcal{M}(H') \neq \emptyset$. Let $M \in \mathcal{M}(H') \setminus \mathcal{M}(H)$ and $M_1, \ldots, M_k$ be the sequence of subgraphs removed by the procedure to obtain H. Since $M \in \mathcal{M}(H') \setminus \mathcal{M}(H)$, we have that $M \notin \mathcal{M}(H)$ and, thus, $M = M_j$ for some $1 \leq j \leq k$ (M is one of the removed subgraphs). Let $\mathcal{U}_0 = \emptyset$ and $\mathcal{U}_i = \mathcal{U}_{i-1} \cup \{M_i\}$, for $1 \leq i \leq k$. Note that $\mathcal{M}(G) \setminus \mathcal{U}_k = \mathcal{M}(H)$ and, given the deletion condition in the procedure, it is clear that $p(M_i, H_{i-1}) < \tau$, for $1 \leq i \leq k$, where H_{i-1} is the subgraph of G induced by the vertices of all subgraphs in $\mathcal{M}(G) \setminus \mathcal{U}_{i-1}$. Since $\mathcal{M}(H') \subseteq \mathcal{M}(G)$ and p is monotone, we also have that $\mathcal{M}(H') \setminus \mathcal{U}_{i-1} \subseteq \mathcal{M}(G) \setminus \mathcal{U}_{i-1}$ and $p(M_i, H'_{i-1}) < \tau$, for $1 \leq i \leq k$, where H'_{i-1} is the subgraph of H_{i-1} induced by the vertices of all subgraphs in $\mathcal{M}(H') \setminus \mathcal{U}_{i-1}$. In particular $p(M, H'_{j-1}) < \tau$ and, thus, M should be removed in the procedure. Hence, if H' was returned, we have that $M \notin \mathcal{M}(H')$ for any $M \in \mathcal{M}(H') \setminus \mathcal{M}(H)$ – an absurd. So, $\mathcal{M}(H') \setminus \mathcal{M}(H) = \emptyset$ and, by an analogous argument, $\mathcal{M}(H) \setminus \mathcal{M}(H') = \emptyset$, i.e., $\mathcal{M}(H) = \mathcal{M}(H')$ and $H = H'$. Therefore, H is unique, independent of the order of subgraph removal and maximal by construction, i.e., 2 holds. ∎

Corollary 2. *Given a monotonic subgraph property function p and $\tau_1, \tau_2 \in \mathbb{R}$ such that $\tau_1 < \tau_2$, the τ_1-core H_1 and the τ_2-core H_2 with respect to p are nested, i.e., H_2 is subgraph of H_1.*

Proof. By Theorem 2 we have that H_1 and H_2 are unique and independent of the order of deletions. Thus, since $\tau_1 < \tau_2$, we may apply the procedure to obtain H_1 and, by continuing the procedure, we may remove more subgraphs to obtain H_2. Therefore, H_2 is a subgraph of H_1. ∎

Although a subgraph property function p is only required to be defined over a set of subgraphs $\mathcal{M}$, the following result holds whenever p is extensible to any set $\mathcal{M}$, namely $p : 2^G \times 2^G \longrightarrow \mathbb{R}$ is well defined.

Proposition 1. *Let G be a graph, p be a monotonic subgraph property function over 2^G, $\tau \in \mathbb{R}$ and $\mathcal{M}, \mathcal{M}' \subseteq 2^G$. If all subgraphs $M' \in \mathcal{M}'$ can be induced by a sequence of subgraphs $M_1, \ldots, M_k \in \mathcal{M}$, i.e., M' is a subgraph induced by $\bigcup_{i=1}^{k} V(M_i)$, then the τ-core H' with respect to p over $\mathcal{M}'$ is a subgraph of the τ-core H with respect to p over $\mathcal{M}$.*

Proof. Since H' is a τ-core with respect to p over $\mathcal{M}'$, there are $M'_1, \ldots, M'_\ell \in \mathcal{M}'$ such that H' is the subgraph induced by $\bigcup_{j=1}^{\ell} V(M'_j)$ and $p(M'_j, H') \geq \tau$, for $1 \leq i \leq \ell$. By hypothesis, each M'_j is a subgraph induced by $\bigcup_{i=1}^{k} V(M_i)$, where $M_1, \ldots, M_k \in \mathcal{M}$. Then, since p is monotone, $p(M_i, H') \geq p(M'_j, H') \geq \tau$, for $1 \leq i \leq k$, and thus $M_1, \ldots, M_k$ are part of the τ-core with respect to p over $\mathcal{M}$. Therefore, all $M_1, \ldots, M_k$, for all M'_j with $1 \leq j \leq \ell$, are subgraphs of H, i.e., H' is subgraph of H. ∎

By Proposition 1, given a suitable subgraph property function, we are able to incrementally build the τ-core by refining the set of subgraphs $\mathcal{M}$. For instance, let p be the subgraph property function

$$p(M, H) = |V(M) \cap V(H)| + |X \cap V(H)|, \tag{6}$$

where $X = \bigcap_{u \in V(M)} N_G(u)$. Note that p is monotone only if we restrict M to cliques. Taking $\mathcal{M}$ as the set of singleton subgraphs, i.e., $\mathcal{M} = \{(\{u\}, \emptyset) \mid u \in V\}$, Eq. (6) is equivalent to Eq. (3) minus one. Thus, given $k \in \mathbb{N}$, a k-core with respect to p over $\mathcal{M}$ is a classical $(k-1)$-core as defined by Seidman. If we take $\mathcal{M}'$ as the set of subgraphs induced by E, i.e., $\mathcal{M}' = \{(\{u, v\}, \{(u, v)\}) \mid (u, v) \in E\}$,

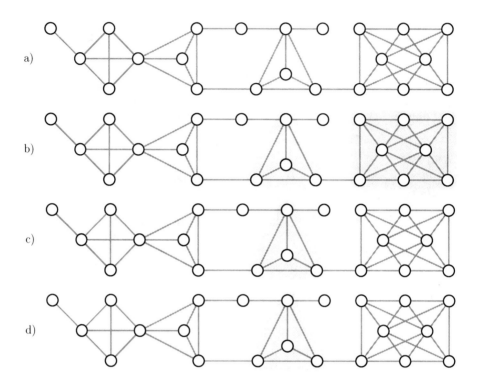

Fig. 2. Graph cores identified using the subgraph property function (6) and different sets of subgraphs $\mathcal{M}$, including the comparison with k-clique percolations. The shadowed vertices are: a) a 4-core with respect to (6) over $\mathcal{K}_1$, *i.e.*, a classical 3-core; b) a 4-core with respect to (6) over $\mathcal{K}_2$, *i.e.*, three classical 4-dense communities; c) a 4-core with respect to (6) over $\mathcal{K}_3$; d) three 4-clique percolation communities. Note that the clique percolations in d) are subgraphs of the core in c), which is subgraph of the core in b), which is also subgraph of the core in a).

the k-cores with respect to p over $\mathcal{M}'$ are precisely the k-dense communities as proposed by Saito *et al.*. Note that, given $k \in \mathbb{N}$, the k-dense community requires that two connected vertices share at least $k-2$ neighbors. In a way analogous to the method proposed by Saito *et al.*, we can compute a k-core with respect to p over $\mathcal{M}$ and, then, we can refine it to obtain a k-dense community by computing a k-core with respect to p over $\mathcal{M}'$. This is a straightforward application of Proposition 1, as illustrated by the two first cases in Fig. 2.

Clearly, we can consider any set of subgraphs with the subgraph property function (6). For instance, given $\ell \in \mathbb{N}$, let $\mathcal{K}_\ell$ be the set of subgraphs of G isomorphic to the clique $K_{\ell'}$ of size ℓ', for all $\ell' \leq \ell$, *i.e.*,

$$\mathcal{K}_\ell(G) = \{H \mid H \text{ is subgraph of } G, H \simeq K_{\ell'} \text{ and } \ell' \leq \ell\}. \tag{7}$$

Note that if we consider $\ell = 1$ or $\ell = 2$, we recover the definitions of classical k-core and k-dense community, respectively. Moreover, for any $k \in \mathbb{N}$, each vertex in the k-core with respect to p over $\mathcal{K}_{k-1}$ belongs to at least one k-clique. This is interesting since it is closely related to the communities found with the clique percolation method [8]. In particular a k-clique percolation community is a subgraph of the k-core with respect to p over $\mathcal{K}_{k-1}$ and, by Proposition 1, it is a subgraph of the k-core with respect to p over $\mathcal{K}_\ell$ for any $\ell < k$ (see Fig. 2). In particular, this establishes a relation of nesting between classical k-cores, k-dense communities and k-clique communities.

As we did for Eq. (3), we can easily extend Eq. (6) to weighted graphs. Given a vertex similarity function σ, the subgraph property function becomes

$$p(M, H) = \sum_{u \in V(M)} \sum_{v \in X \cap V(H)} \sigma(u, v), \tag{8}$$

where $X = \bigcap_{u \in V(M)} N_G(u)$. Note that p is monotone only if the weights are equal to 1.0, otherwise the second monotonocity contidion may not hold. Taking σ as the weight function w and considering $\mathcal{M} = \mathcal{K}_2$, Eq. (8) is the natural extension of the k-dense notion to weighted graphs leading to the notion of τ-dense community with $\tau \in \mathbb{R}$.

The Corollary 2 (or 1 in the simpler case) ensures that, given a monotonic subgraph property function p, we can built a hierarchy of nested cores by considering different values of τ. This is interesting since, by ranging over different values of τ, we get a hierarchy of cores.

4   Discussion and Applications

In this paper we propose fully generalized cores, which extend several core definitions proposed in the literature under a common framework. Moreover, we discuss a greedy approach to solve the problem of identifying fully generalized cores. The complexity of this approach is clearly dependent on subgraph property functions, which may be computationally costly. Although for some subgraph property functions this problem can be stated as graphic matroid [21,22], it remains to be seen under which formal conditions this combinatorial problem becomes a matroid.

In what concerns interesting and desirable properties, there are other related approaches to core enumeration. Recently Xu et al. [20] implicitly proposed the following alternative definition of core. Given the similarity function (1), $n \in \mathbb{N}$ and $\varepsilon > 0$, we say that $(u, v) \in E$ is a core edge if $\sigma(u, v) \geq \varepsilon$, and that $u \in V$ is a core vertex if $|\{v \in N(u) \mid \sigma(u, v) \geq \varepsilon\}| \geq n$. Then, a set of vertices $C \subseteq V$ is a core in G if all $u \in C$ is a core vertex and if, for all $u, v \in C$, there is a connecting path composed only of core edges. The parameter n is the main difference with respect to the core enumeration approaches discussed in this paper. Given $n \in \mathbb{N}$, we compute the ε-core H with respect to the property function (5), but we further filter it by leaving just the vertices $u \in V$ such that $|\{v \in H \mid \sigma(u, v) \geq \varepsilon\}| \geq n$. Thus, although the definition of core proposed by Xu *et al.* is related to the notion of generalized core, it introduces an extra degree of freedom that is interesting if we require higher resolutions.

There are several interesting applications for fully generalized cores. Here, we briefly discuss two of them. As discussed before, an application is the detection of densely connected subgraphs within graph clustering methods. Given a core, we can take each connected component as a seed set and apply well known local partition methods [25,26,27]. Note that by using the approach described in this paper, we can get a hierarchy of cores and, thus, we are able to get a hierarchical clustering. There are several alternatives for hierarchical clustering and local optimization. For instance, Lancichinetti *et al.* [28] proposed a multiresolution method that optimizes a local fitness score by adding and removing vertices to increase the fitness score, following an approach like the one proposed by Blondel *et al.* [29]. These are equivalent to the approaches based on ranking, where each vertex constitutes a core or seed set. The main issue with these simpler approaches is that there is not any guarantee about their effectiveness. On the other hand, local ranking based on, *e.g.*, the heat kernel has supporting results both with respect to local optimization complexity and clustering quality [27]. These approaches allow also for the detection of vertices that appear in multiple clusters, *i.e.*, overlapping clusterings. Note also the ability to obtain local clusterings, in particular when we do not know all the graph. This problem is partially addressed by the local optimization or local clustering techniques. But an important issue remains: what happens if the seed set is composed by vertices already within an overlap? If we just use a standard local clustering approach, we will obtain just a big cluster composed of several smaller and overlapping clusters. By partially exploring the neighborhood of the seed set, by enumerating the cores, and by applying local clustering to the obtained seed sets, we can detect the smaller and overlapping clusters.

A second application is the detection of complex network motifs, which we already mentioned. Given a set of motifs or graphlets, we can enumerate the cores composed only by vertices belonging to these motifs or graphlets. The main task becomes defining a suitable subgraph property function. The resulting cores can then be statically evaluated, identifying possible mesoscale network motifs. This is of high importance since enumerating and evaluating motifs or graphlets with a reasonable size is computationally demanding. Unlike graphlets, network motifs may not be induced subgraphs and, thus, we may want to consider the merging of motifs instead of vertex induced subgraphs in our definition of fully generalized cores. The results presented herein remain valid.

References

1. Seidman, S.B.: Network structure and minimum degree. Social Networks 5(3), 269–287 (1983)
2. Wasserman, S., Faust, K.: Social network analysis: Methods and applications. Cambridge University Press, Cambridge (1994)
3. Batagelj, V., Zaveršnik, M.: Generalized cores. arXiv:cs/0202039 (2002)

4. Leskovec, J., Lang, K.J., Dasgupta, A., Mahoney, M.W.: Community structure in large networks: Natural cluster sizes and the absence of large well-define clusters. arXiv:0810.1355 (2008)
5. Wei, F., Qian, W., Wang, C., Zhou, A.: Detecting Overlapping Community Structures in Networks. World Wide Web 12(2), 235–261 (2009)
6. Schloegel, K., Karypis, G., Kumar, V.: Graph partitioning for high-performance scientific simulations. Morgan Kaufmann Publishers, Inc., San Francisco (2003)
7. Abou-Rjeili, A., Karypis, G.: Multilevel algorithms for partitioning power-law graphs. In: IEEE International Parallel & Distributed Processing Symposium, p. 10. IEEE, Los Alamitos (2006)
8. Palla, G., Derényi, I., Farkas, I., Vicsek, T.: Uncovering the overlapping community structure of complex networks in nature and society. Nature 435, 814–818 (2005)
9. Farkas, I., Ábel, D., Palla, G., Vicsek, T.: Weighted network modules. New J. Physics 9(6), 180 (2007)
10. Shen, H., Cheng, X., Cai, K., Hu, M.B.: Detect overlapping and hierarchical community structure in networks. Physica A: Statistical Mechanics and its Applications 388(8), 1706–1712 (2009)
11. Saito, K., Yamada, T., Kazama, K.: Extracting Communities from Complex Networks by the k-dense Method. IEICE Transactions on Fundamentals of Electronics, Communications and Computer Sciences 91, 3304–3311 (2008)
12. Milo, R., Shen-Orr, S., Itzkovitz, S., Kashtan, N., Chklovskii, D., Alon, U.: Network Motifs: Simple Building Blocks of Ccomplex Networks. Science 298, 824–827 (2002)
13. Saito, R., Suzuki, H., Hayashizaki, Y.: Construction of reliable protein-protein interaction networks with a new interaction generality measure. Bioinformatics 19(6), 756–763 (2002)
14. Albert, I., Albert, R.: Conserved network motifs allow protein-protein interaction prediction. Bioinformatics 20(18), 3346–3352 (2004)
15. Kashtan, N., Itzkovitz, S., Milo, R., Alon, U.: Efficient sampling algorithm for estimating subgraph concentrations and detecting network motifs. Bioinformatics 20(11), 1746–1758 (2004)
16. Kuramochi, M., Karypis, G.: An efficient algorithm for discovering frequent subgraphs. IEEE Transactions on Knowledge and Data Engineering 16(9), 1038–1051 (2004)
17. Milenković, T., Lai, J., Pržulj, N.: Graphcrunch: a tool for large network analyses. BMC Bioinformatics 9(1), 70 (2008)
18. Jaccard, P.: Distribution de la flore alpine dans le Bassin des Dranses et dans quelques regions voisines. Bull. Soc. Vaud. Sci. Nat. 37, 241–272 (1901)
19. Tanimoto, T.T.: IBM Internal Report November 17th. Technical report, IBM (1957)
20. Xu, X., Yuruk, N., Feng, Z., Schweiger, T.A.J.: Scan: a structural clustering algorithm for networks. In: SIGKDD, pp. 824–833. ACM, New York (2007)
21. Whitney, H.: On the abstract properties of linear dependence. American Journal of Mathematics 57(3), 509–533 (1935)
22. Tutte, W.T.: Lectures on matroids. J. Res. Nat. Bur. Stand. B 69, 1–47 (1965)
23. Kruskal, J.B.: On the shortest spanning subtree of a graph and the traveling salesman problem. Proceedings of the AMS 7(1), 48–50 (1956)
24. Borůvka, O.: On a minimal problem. Prace Moraské Pridovedecké Spolecnosti 3 (1926)

25. Spielman, D.A., Teng, S.H.: A local clustering algorithm for massive graphs and its application to nearly-linear time graph partitioning. arXiv.org:0809.3232 (2008)
26. Andersen, R., Lang, K.J.: Communities from seed sets. In: WWW, pp. 223–232. ACM, New York (2006)
27. Chung, F.: The heat kernel as the pagerank of a graph. PNAS 104(50), 19735–19740 (2007)
28. Lancichinetti, A., Fortunato, S., Kertész, J.: Detecting the overlapping and hierarchical community structure in complex networks. New J. Physics 11, 33015 (2009)
29. Blondel, V.D., Guillaume, J.L., Lambiotte, R., Lefebvre, E.: Fast unfolding of communities in large networks. Journal of Statistical Mechanics: Theory and Experiment, 10008 (2008)

Connectivity Criteria for Ranking Network Nodes

Jaime Cohen[1], Elias P. Duarte Jr.[2], and Jonatan Schroeder[3]

[1] Federal University of Paraná and State University of Ponta Grossa
jaimecohen@gmail.com
[2] Federal University of Paraná
elias@inf.ufpr.br
[3] University of British Columbia
jonatanschroeder@gmail.com

Abstract. In this work we introduce a new quantitative criteria for assessing the connectivity of nodes based on the well known concept of edge-connectivity. We call the new criteria the *connectivity numbers* of a node. They consist of a hierarchy of measures that starts with a local measure that progressively becomes a global connectivity measure of the network. We show that the connectivity numbers can be computed in polynomial time. Experimental results are described showing how the proposed approach compares to other well known concepts involving connectivity and centrality of network nodes in real and synthetic networks.

1 Introduction

Network topology analysis allows the assessment of several important network properties, such as connectivity, dynamics, and resiliency. For instance, the robustness of many computer applications can be improved when the underlying topology is fully understood [1,2]. Important problems such as routing and resource placement are among those that can benefit from this knowledge.

The present work introduces a set of quantitative criteria for ranking nodes and assessing the reliability of networks in terms of connectivity. These criteria are based on the well known concept of edge-connectivity and aim at giving a precise reply to the frequently asked question: given a network topology, how can we rank nodes in terms of their connectivity?

To rank network nodes with respect to their connectivity, we define a measure called *connectivity number*, denoted by i-*connectivity*(v) or $\#C_i(v)$, for short, that corresponds to the connectivity of the node v with respect to sets of nodes of size at least i. For small values of i, the value of $\#C_i(v)$ corresponds to a local measure of connectivity. In particular, $\#C_1(v)$ is defined as the degree of node v. Large values of i correspond to global measures of connectivity and for i equal to the number of nodes of the network, the connectivity number corresponds to the size of a global minimum edge-cut.

Experimental results are reported. We compare the proposed criteria with other criteria such as node degree and eccentricity. We show, for instance, that in real networks from different domains, the ranking of the nodes of large degree in order of connectivity numbers differs from the ranking in order of degree. We also show that nodes with high connectivity numbers often have low eccentricity (i.e. they are central). The

L. da F. Costa et al. (Eds.): CompleNet 2010, CCIS 116, pp. 35–45, 2011.

criteria were also evaluated by the sequential removal of network nodes for determining how aggressively faulty nodes can disconnect a network. This experiment shows that the criteria can be used to determine a set of core nodes that, if removed, rapidly increases the number of disconnected nodes.

The rest of this paper is organized as follows. The next section presents related work. Then, the proposed connectivity criteria are defined followed by the specification of a polynomial-time algorithm to compute the criteria. Experimental results comparing the proposed criteria with other measures of connectivity are presented next. The last section concludes the paper.

2   Related Work

The local properties of a graph such as node degree do not necessarily describe its relevant properties for a particular application. As pointed out in [3], local and global properties of a graph are "separable". For instance, graphs with different properties as tertiary trees, two-dimensional grids and random graphs with degree four have the same degree distribution but very different global structure. On the other hand, trees may differ significantly in terms of degree distribution but share the same hierarchical structure. Shavitt and Singer [4] show that neither the node degree nor its centrality are good measures for classifying the importance of nodes for efficient routing in the Internet. The authors define two new metrics in an attempt to better characterize nodes in terms of their importance for routing.

Wuchty and Stadler [5] study common centrality properties of graphs in the context of complex biological networks. The authors argue that the vertex degree is a local measure that only provides relevant information when the graph is generated with a known probabilistic distribution. They evaluate the concepts of eccentricity, status and the centroid value of a vertex and compare them with vertex degree. They show that those measures often correlate to the degree, but that they differ in exposing topological properties of the network that is not captured by the degree.

Palmer et al. [6] applies the notion of node eccentricity to evaluate the topology of backbones of the Internet. They define the effective eccentricity of a node and apply it to the study of the robustness of a sample of the Internet topology. They show that the effective eccentricity is useful in classifying nodes in terms of importance for the reliability of the network, as the failure of nodes with low eccentricity quickly decreases the connectivity of the network. Even though the removal of nodes in non-increasing order of degree is one of the most aggressive ways to disconnect an Internet like topology, the authors argue that since the eccentricity measures other node properties not captured by the degree, it is an interesting measure to locate critical nodes of networks.

A routing scheme based on the connectivity criteria $\#C_2(v)$ that consists of using alternative routes called *detours* to bypass network failures was proposed in [7].

The study of network measurements is a broad area of study that has been evolving for several decades with contributions by researchers from mathematics, computer science, physics, statistics, sociology, economy and biology. Numerous concepts related to centrality of networks have been proposed. Some of them are based on distances and paths, such as average distances and betweenness, others are based on clustering and cycles, on spectral methods, on hierarchical properties, among others. The interested reader will find good reviews on the topic in [8] and [9].

3   The Proposed Connectivity Criteria

In this section we describe the connectivity numbers. The network topology is represented as an undirected graph $G = (V, E)$ where V is a set of vertices representing the network nodes, and E is a set of edges representing the network links. Although only unweighted graphs are treated in this paper, all definitions and algorithms can be adapted to weighted graphs by considering the weight of the cuts instead of their cardinality.

Our goal is to find a good measure for the connectivity of a node with respect to the network. The most obvious such measure is the degree of the vertex. However, in many applications, the degree of a vertex is not an ideal measure of connectivity. For example, consider the problem of finding core routers given the topology of the network. A node with high degree may be at the periphery of the network while nodes with low eccentricity (central nodes) may have low degree and both may correspond to unimportant nodes.

In order to circumvent the problem, we propose a novel approach for measuring the connectivity of nodes with respect to the network. The concept was built upon a well known definition of edge-connectivity between pairs of nodes of a graph.

We start with a few definitions. Let $G = (V, E)$ be an undirected graph. Let $X \subseteq V$. We denote by $\delta(X)$ the set of edges $\{u, v\}$ such that $u \in X$ and $v \in V \setminus X$. We call a set of edges $C \subseteq E, C \neq \emptyset$, a *cut* if $C = \delta(X)$ for some $X \subseteq V$. A cut $C = \delta(X)$ *separates* two vertices s and t if $s \in X$ and $t \in V \setminus X$. A *minimum s-t-cut* is a cut of minimum cardinality that separates s from t.

In unweighted graphs, the dual of a cut separating two vertices s and t is a collection of edge-disjoint paths connecting s and t. The size of a minimum cut that separates s from t equals the maximum number of edge-disjoint paths connecting s and t [12].

Now, an important definition is given:

Definition 1. *Let $G = (V, E)$ be an undirected graph. Consider a set of vertices $X \subseteq V$. The* edge-connectivity *of X with respect to G is the size of the minimum cut that separates any pair of vertices in X.*

It is important to note that the edge-connectivity of a set of nodes X is different from the edge-connectivity of the subgraph induced by X. The reason is that the minimum-cuts are taken from G and not from the subgraph induced by X. The subgraph induced by X may not have any edges and still X may have a high edge-connectivity. See figure 1 for an illustration. The set $X = \{a, b, c, d\}$ has edge-connectivity 3 even though the graph induced by X has no edges.

Now we define the connectivity criteria.

Definition 2. *The* connectivity number with index i *of a node v, denoted by* i-connectivity(v) *or* $\#C_i(v)$ *for short, is the maximum edge-connectivity of a set $X \subseteq V$ satisfying*

   i. $v \in X$, *and*
   ii. $|X| \geq i$

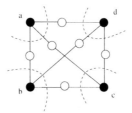

Fig. 1. Set $X = \{a,b,c,d\}$ has edge-connectivity 3 with respect to the graph. Dashed lines represent minimum cuts separating vertices of X.

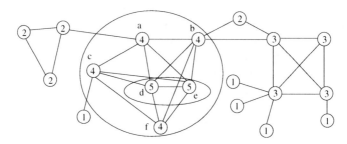

Fig. 2. Connectivity criteria $\#C_2(v)$. Circles show some components of maximum connectivity.

Figure 2 shows an example of the connectivity numbers. The numbers associated with nodes are the connectivity numbers $\#C_2(v)$. The circled areas identify components of maximum connectivity. To compute the 2-connectivities, we look for the maximum edge-connectivity of sets containing 2 nodes. For example, $\#C_2(a) = 4$ because nodes a and b cannot be separated by a cut with less than 4 edges and the edge-connectivity between a and any other node is at most 4. To compute 3-connectivities, only sets with at least 3 nodes are considered. For example, $\#C_3(d) = 4$ because any set containing d with at least three nodes has edge-connectivity at most 4. Note, however, that $\#C_2(d) = 5$.

By convention, $\#C_1(v)$ is defined to be the degree of v and $\#C_n(v)$, where $n = |V|$, is, by the definition, the size of the global minimum edge-cut. The i-connectivities are a unified measure that, as i increases from 1 to $|V|$, $\#C_i(v)$ ranges from a local property to a global property of the network. The following results are straightforward:

Lemma 1. *Let $G = (V,E)$ be a graph and $v \in V$. Then,*

$$degree(v) = \#C_1(v) \geq \#C_2(v) \geq \#C_3(v) \geq \ldots \geq \#C_n(v) = \lambda(G)$$

where $\lambda(G)$ is the size of a minimum global edge cut of G.

Lemma 2. *Let $G = (V,E)$ be a graph. If V' is a subset of V of maximum size such that $v \in V'$ and V' has edge-connectivity $\#C_i(v)$, then V' is unique.*

The proofs of the lemmas above are straightforward.

Let us consider the 2-connectivities. The following lemma is the key to the computation of $\#C_2(v)$:

Lemma 3. *Given a graph $G = (V,E)$ and a vertex $v \in V$, the value of $\#C_2(v)$ is equal to the maximum cardinality among all minimum cuts separating v from any other vertex of G.*

The proof of this lemma follows from the definition of the connectivity numbers. In the next section we show an algorithm to compute i-connectivity(v).

4 An Algorithm for Computing the Connectivity Numbers

First we show how $\#C_2(v)$ can be computed and then we generalize the result for $\#C_i(v)$ for all i. The connectivity number $\#C_2(v)$ of a vertex v is equal to the largest number of edges in a minimum cut that separates v from some other vertex and therefore we can compute $\#C_2(v)$, for all $v \in V$, with an algorithm for the all pairs minimum cut problem.

A *cut tree* is a compact representation of the minimum cuts that separates every pair of vertices in a graph [10]. We show below that all connectivity criteria described in the previous section can be computed efficiently given the cut tree of the graph.

A *cut tree* T of a graph G is defined as a weighted tree such that 1) the vertices of G correspond to the vertices of the tree T. 2) The size of a minimum cut between two vertices of G is given by the minimum weight of an edge in the unique path between the two corresponding vertices in T. 3) A minimum cut between two nodes of G is given by removing the edge that attains the minimum value described in (2) and taking the two vertex sets induced by the two trees formed by removing that edge from T.

Figure 3 shows a cut tree associated with a graph. To determine the number of edges in the minimum cut we look for the minimum weight of an edge that belongs to the unique path between the two corresponding vertices in the cut tree.

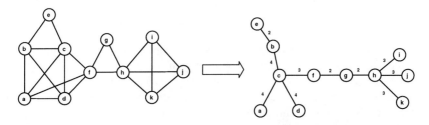

Fig. 3. A graph and the corresponding cut-tree

The third condition that defines a cut tree is not strictly necessary for our needs and in fact a tree that satisfies the first two properties has been named a *flow equivalent tree*, since the property that the pairwise maximum flow between vertices has the same values in the tree and in the associated graph [11]. Known time bounds for constructing flow equivalent trees are the same as those for cut trees.

Having computed a cut tree for the graph, we can find the connectivity numbers as described below.

Lemma 4. *For a graph G, a cut tree T of G and vertex v, the value of $\#C_2(v)$ is equal to the largest weight of an edge incident to v in T.*

Proof. By definition, the cardinality of a minimum cut separating v from another vertex, say u, corresponds to the minimum weight of an edge in the unique path from u to v in T. Furthermore, this value must be less than or equal to the weight of an edge that is incident to v in T. Therefore, the maximum capacity of any minimum cut separating v from other vertices must be equal to the weight of some edge of T incident to v, more precisely, the one with the maximum weight.   □

Lemma 4 shows that for each node v, $\#C_2(v)$ can be computed in linear time from the cut tree by considering the edge incident to v in T with the largest weight.

The idea described above can be generalized to find $\#C_i(v)$ for all $i, 2 \le i \le n$. The generalized algorithm is shown below:

Algorithm for computing $\#C_i(v)$
Input: A graph $G = (V,E)$, $v \in V$ and an integer i, $1 \le i \le |V|$
Output: $\#C_i(v)$

```
if i = 1 then return degree(v)
Let T = (V, E_T, w) be a cut tree of G;
C ← {v}
cn ← ∞;
while (|C| < i) do
    let e = (x, y) ∈ E_T be an edge with x ∈ C and y ∈ V \ C of maximum weight;
    C ← C ∪ {y};
    if w(e) < cn then cn ← w(e);
end_while
return #C_i(v) = cn;
```

Let n be the number of vertices in V and m the number of edges in E. The construction of the cut tree dominates the time complexity of the algorithm. Given the cut tree, the rest of the algorithm can be implemented to run in $O(n.log(n))$ if steps 4,6 and 8 are implemented using a binary heap. In fact, all connectivity numbers can be computed with the same time bound with an appropriate transversal of the cut tree. Below, we discuss the complexity to build a cut tree.

In order to compute the minimum cut between a pair of vertices in G, Dinits' algorithm runs in time $\mathcal{O}(m^{\frac{3}{2}})$ (or $\mathcal{O}(n\sqrt{n})$ for sparse graphs) [12]. Two algorithms for constructing a cut tree are known, one by R.E. Gomory and T.C. HU [10] and another by D. Gusfield [11]. Both algorithms computes minimum cuts between pair of vertices $n - 1$ times. Thus, the complexity of these algorithms are $\mathcal{O}(nm \cdot \sqrt{m})$ (or $n^2 \cdot \sqrt{n}$ on sparse graphs). Given the cut tree, $\#C_i(v)$, for all $v \in V$, can be computed in subquadratic time on the number of nodes. In unweighed graphs, the cut tree can be found in $\mathcal{O}(nm)$ with a recent algorithm by Hariharan, R. et al. [13].

5   Experimental Evaluation

In this section we present experimental results showing basic properties of the connectivity numbers and comparing various measures of node centrality in real and synthetic networks.

5.1   The i-Connectivity of Real Networks

We collected a set of 9 real networks from different domains. Networks *CA-AstroPh, CA-CondMat, CA-HepPh, CA-GrQc* [14] and *geocomp* [15] are research collaboration networks, *powergrid* [16] is a power grid network, p2p-Gnutella04 and p2p-Gnutella25 [14] are P2P networks and yeast [17] is a protein-protein interaction network. Table 1 shows the sizes of the theses networks. Only the maximum connected components of the underlying networks were used.

Table 1. Sizes of the 9 real graphs used in the correlation experiments

	CA-AstroPh	CA-CondMat	CA-GrQc	CA-HepPh	geocomp	p2p-Gnut04	p2p-Gnut25	yeast	powergrid
$|V|$	17903	21363	4158	11204	3621	10876	22663	2221	4941
$|E|$	393944	182572	26844	235238	9461	39994	54693	6602	6594

Table 2 shows basic statistics of the real networks used in the experiments. It shows the minimum, the maximum and the average degree. The same parameters are shown for the 2- and 3-connectivities.

Table 2. Comparison of the degree and the 2- and 3-connectivity of nine real networks. The table shows minimum, maximum and mean values.

Network	Minimum			Maximum			Mean		
	Degree	$\#C_2(v)$	$\#C_3(v)$	Degree	$\#C_2(v)$	$\#C_3(v)$	Degree	$\#C_2(v)$	$\#C_3(v)$
CA-AstroPh	2	2	2	1008	854	838	44	43.8	43.7
CA-CondMat	2	2	2	558	504	394	17.1	16.6	16.6
CA-GrQc	2	2	2	162	154	150	12.9	12.1	12.1
CA-HepPh	2	2	2	982	972	964	42	41.6	41.5
geocomp	1	1	1	102	431	359	5.2	9.9	9.7
p2p-Gnut25	1	1	1	66	57	45	4.8	4.4	4.4
p2p-Gnut04	1	1	1	103	81	65	7.4	7.1	7.1
yeast	1	1	1	64	59	55	5.9	5.6	5.6
powergrid	1	1	1	19	12	11	2.7	2.3	2.3

Ranking Nodes in Real Networks. To show that the rankings of nodes in order of degree and in order of i-connectivity differ, we computed rank correlations between node degree and i-connectivity of the first 10 and 100 nodes with largest degrees. We used *Kendall's rank correlation* which values are +1 for sequences with the same order and -1 for those in reverse.

Results are shown in Table 3. The networks with the highest correlations between degree and connectivity numbers are the first four collaboration networks. The networks that present the lowest correlations are *powergrid*, *geocomp* and *p2p-gnuttella25*, in this order, being as low as 0.14 for the *powergrid* and 0.48 for *geocomp*.

Table 3. Rank correlation between the degree and the connectivity numbers of nodes with the 10 and 100 largest degree

| | Kendall Rank Correlation | | | |
| | Largest 10 | | Largest 100 | |
Network	$\#C_2(v)$	$\#C_3(v)$	$\#C_2(v)$	$\#C_3(v)$
CA-AstroPh	0.99	0.97	1	1
CA-CondMat	0.98	0.95	0.94	0.94
CA-GrQc	0.81	0.83	0.94	0.93
CA-HepPh	0.91	0.88	0.99	0.99
geocomp	0.75	0.72	0.82	0.82
p2p-Gnut25	0.54	0.51	0.64	0.64
p2p-Gnut04	0.78	0.72	0.92	0.92
yeast	0.8	0.77	0.85	0.85
powergrid	0.2	0.14	0.35	0.33

5.2   Synthetic Networks

The next experiments were done in synthetic graphs generated using the algorithm by Bu and Towsley [18]. The parameter values used by this algorithm are p, the probability of creating a new node in each step, instead of only an edge, and β, the preference to high-degree nodes for new edges. We used the values $p = 0.4695$ and $\beta = 0.6447$ that are the same found by Bu and Towsley to generate topologies similar to the Internet.

Experiments were performed in order to compare our proposed criteria and other common criteria for assessing the centrality of network nodes. The experiments show that $\#C_i(v)$ captures different aspects of network topology than others measures. We compared the node degree, $\#C_i(v)$, with $i \in \{2,3\}$, eccentricity and effective eccentricity. All results are averages for 100 different graphs with 1000 nodes.

Node Degree and *i*-Connectivity. Initially, a comparison between node degree and the 2-connectivity was performed. The results of this comparison are shown in figure 4. In each graph, the diagonal line corresponds to the identity function and helps visualizing the difference between the degree and the *i*-connectivity of the nodes. Some crosses may correspond to multiple occurrences of nodes with the same degree and the same *i*-connectivity.

We observe that the higher the degree of a node, the higher is the expected difference between the *i*-connectivity and the degree. Most importantly, *nodes with the higher degrees are not necessarily the ones with the higher i-connectivities*. For example, a node with degree equal to 262 has 2-connectivity equal to 100, while a node in the same graph with degree equal to 140 has 2-connectivity greater than 100.

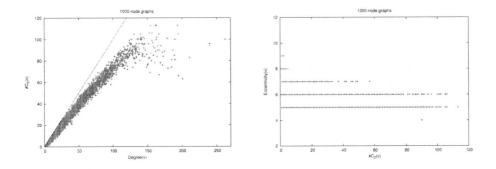

Fig. 4. Comparison between node degree and *i*-connectivity

Fig. 5. Comparison between eccentricity and *i*-connectivity

Eccentricity and *i*-Connectivity. The *eccentricity* of a node v is defined as the largest distance between v and any other node in the graph. The *effective eccentricity* is the number of hops needed to reach at least 90% of the nodes in the network. These criteria distinguish the central nodes and the peripheral ones.

We compared 2-connectivity, eccentricity and effective eccentricity of nodes. The results are shown in figures 5 and 6, respectively for the eccentricity and the effective eccentricity.

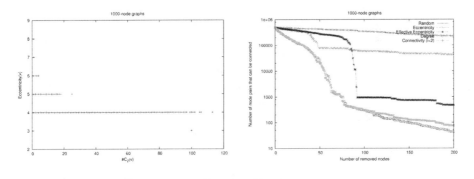

Fig. 6. Comparison between effective eccentricity and 2-connectivity

Fig. 7. Number of node pairs that are connected after sequential node removal

The results for the comparison with eccentricity show that nodes with high connectivity are usually closer to the center of the graph, that is, they have lower eccentricity, while nodes with high eccentricity usually have a lower connectivity. Besides, there are many nodes with low connectivity and low eccentricity, *showing that eccentricity alone is not always a reliable criterion for assessing the centrality a node*. The results are similar for effective eccentricity.

Sequential Node Removal. The execution of a sequential node removal shows how aggressively one can disconnect a network by removing nodes. The number of connected pairs of nodes is computed after the cumulative removal of nodes ordered by the several criteria. It is well known that power-law networks are vulnerable to attacks to nodes with high degree.

We evaluated the number of disconnected pairs of nodes when nodes were removed in random order, non-increasing order of 2-connectivity, node degree and in non-decreasing order of their eccentricity and effective eccentricity.

Figure 7 shows the results of the experiment. In all of the graphs, the x-axis represents the number of removed nodes and the y-axis represents the number of connected pairs of nodes. Removal of only the first 20% of the nodes were simulated in all cases.

The results found in this experiment show that the selection of nodes with high connectivity and with high degree have similar results, with a small advantage to node degree. The results also show that both 2-connectivity and degree disconnect the network faster than the selections based on eccentricity.

6   Conclusions

This paper introduced connectivity criteria for ranking network nodes. The criteria can be used to identify core nodes in networks. An exact polynomial time algorithm for computing the criteria was given.

We showed that in some real networks, the ranking of nodes with large degree is different when the ordering criteria is the degree or the i-connectivity.

Experimental results in synthetic graphs revealed the existence of nodes with high degree, but relatively low i-connectivity in power law networks. We showed that nodes with high connectivity numbers have low eccentricity, supporting the idea of using the criteria to locate core nodes of networks. On the other hand, the same networks have nodes with low connectivity but low eccentricity, showing that eccentricity alone is not enough for locating well connected nodes.

Future work includes the study of analogous criteria for vertex-connectivity instead of edge-connectivity and further analysis of real complex networks through the lens of the i-connectivity.

Acknowledgements. This work was partially supported by grant 304013/2009-9 from the Brazilian Research Agency (CNPq). Jaime Cohen was on a paid leave of absence from UEPG to conclude his Ph.D. and he was also supported by a Fundação Araucária/ SETI fellowship under Order No. 16/2008.

References

1. Han, J., Watson, D., Jahanian, F.: An Experimental Study of Internet Path Diversity. IEEE Transactions on Dependable and Secure Computing 3(4) (2006)
2. Apostolopoulos, J.G., Trott, M.D.: Path Diversity for Enhanced Media Streaming. IEEE Communications Magazine 42(8) (2004)

3. Tangmunarunkit, H., Govindan, R., Jamin, S., Shenker, S., Willinger, W.: Network Topology Generators: Degree-Based vs. Structural. ACM SIGCOMM Computer Communication Review 32(4) (2002)
4. Shavitt, Y., Singer, Y.: Beyond Centrality - Classifying Topological Significance using Backup Efficiency and Alternative Paths. New Journal of Physics, Focus Issue: Complex Networked Systems: Theory and Applications (2007)
5. Wuchty, S., Stadler, P.F.: Centers of Complex Networks. Journal of Theoretical Biology 223(1) (2003)
6. Palmer, C.R., Siganos, G., Faloutsos, M., Faloutsos, C., Gibbons, P.: The Connectivity and Fault-Tolerance of the Internet Topology. In: Proc. Workshop on Network-Related Data Management (2001)
7. Duarte Jr., E.P., Santini, R., Cohen, J.: Delivering Packets During the Routing Convergence Latency Interval through Highly Connected Detours. In: Proc. IEEE Int'l Conf. Dependable Systems and Networks (2004)
8. da, L., Costa, F., Rodrigues, F.A., Travieso, G., Villas Boas, P.R.: Characterization of complex networks: A survey of measurements. Advances in Physics 56(1), 167–242 (2007)
9. Brandes, U., Erlebach, T. (eds.): Network Analysis: Methodological Foundations. Lecture Notes in Computer Science / Theoretical Computer Science and General Issues. Springer, Heidelberg (2005)
10. Gomory, R.E., Hu, T.C.: Multi-Terminal Network Flows. SIAM Journal on Applied Mathematics 9 (1961)
11. Gusfield, D.: Very Simple Method for All Pairs Network Flow Analysis. SIAM Journal on Computing 19(1) (1990)
12. Nagamochi, H., Ibaraki, T.: Algorithmic Aspects of Graph Connectivity. Cambridge University Press, Cambridge (2008)
13. Hariharan, R., Kavitha, T., Panigrahi, D., Bhalgat, A.: An (mn) Gomory-Hu Tree Construction Algorithm for Unweighted Graphs. In: Proc. 39th Annual ACM Symposium on Theory of Computing, STOC (2007)
14. Leskovec, J., Kleinberg, J., Faloutsos, C.: Graph Evolution: Densification and Shrinking Diameters. ACM Transactions on Knowledge Discovery from Data (ACM TKDD) 1(1) (2007)
15. Batagelj, V., Mrvar, A.: Pajek datasets (2006)
16. Watts, D.J., Strogatz, S.H.: "Collective dynamics of 'small-world' networks. Nature 393, 440–442 (1998)
17. Sun, S., Ling, L., Zhang, N., Li, G., Chen, R.: Topological structure analysis of the protein-protein interaction network in budding yeast. Nucleic Acids Research 31(9) (2003)
18. Bu, T., Towsley, D.F.: On Distinguishing between Internet Power Law Topology Generators. In: Proc. IEEE INFOCOM (2002)

Structure-Dynamics Interplay in Directed Complex Networks with Border Effects

Lucas Antiqueira and Luciano da Fontoura Costa

Institute of Physics at São Carlos, University of São Paulo, São Carlos, Brazil
lantiq@ursa.ifsc.usp.br, luciano@ifsc.usp.br

Abstract. Despite the large number of structural and dynamical properties investigated on complex networks, understanding their interrelationships is also of substantial importance to advance our knowledge on the organizing principles underlying such structures. We use a novel approach to study structure-dynamics correlations where the nodes of directed complex networks were partitioned into border and non-border by using the so-called diversity measurement. The network topology is characterized by the node degree, the most direct indicator of node connectivity, while the dynamics is related to the steady-state random walker occupation probability (called here *node activity*). Correlations between degree and activity were then analyzed inside and outside the border, separately. The obtained results showed that the intricate correlations found in the macaque cortex and in a WWW subgraph are in fact composed of two separate correlations of in-degree against the activity occurring inside and outside the border. These findings pave the way to investigations of possibly similar behavior in other directed complex networks.

Keywords: complex networks, random walk, network border.

1 Introduction

Complex networks are often studied through the analysis of their structural or dynamical properties [4,12]. The structure is related to how the connectivity is organized among nodes; for instance, interesting structural properties are the well-known power-law degree distribution [2] and the small-world effect [21]. On the other hand, the dynamics is based on time-dependent processes taking place on the network (e.g. transport [18] and synchronization [1]). Another trend in network research is the investigation of the interplay between structure and dynamics, where the purpose is to understand how these properties are interrelated, i.e. which specific topological properties lead to a particular dynamical behavior, and vice versa. Investigations on dynamics previously focused on Euclidean networks – both regular and topologically disordered – used as structural models in condensed matter [3], where the topology is dictated by physical properties of the material and often is known, e.g. in the case of lattice models. Recent developments started to investigate dynamical processes on more complex topologies exhibiting a larger number of different scenarios in dynamical behavior, as compared to Euclidean and in particular regular networks (e.g. epidemiological processes on social networks and synchronization on brain networks [3, 4, 12]), leading to intricate structure-dynamics relationships.

L. da F. Costa et al. (Eds.): CompleNet 2010, CCIS 116, pp. 46–56, 2011.

In this context, diffusion processes have a special importance given their wide range of applicability and the existence of different associated models (e.g. the random walk model [15] and Gray-Scott reaction-diffusion model of nonlinear dynamics [14]). In particular, the random walk is a well-known model of dynamics on networks (or graphs [10, 13, 16, 3]), whose variants include the self-avoiding, preferential, and the reinforced random walks. For the simple random walk, where the walker is allowed to visit only neighboring nodes with probability independent of the walking history, it is known that structure and dynamics are perfectly correlated on non-directed networks: the steady-state node occupation probability $\pi(i)$ of a node i can be calculated directly from the number of edges attached to it (i.e. $\pi(i) = k(i)/\sum_{j=1}^{N} k(j)$, where $k(i)$ is the degree of node i and N is the total number of nodes in the network) [10, 13]. For the case of networks with directed edges, a sufficient condition for this perfect correlation to happen is that all nodes have balanced in- and out-degrees, i.e. for each node, the number of incoming links has to be equal to the number of outgoing links [8]. Other cases of directed networks, which are in fact the most common ones, tend to show intricate correlations between the random walk steady-state occupation probability and in-/out-degrees. Examples include the WWW and the *C. elegans* synaptic network [8], as well as the cat thalamocortical network [7].

In this work, we considered a novel approach to investigate structure-dynamics correlations in directed networks: we employed the concept of node diversity [20] to partition networks into border and non-border nodes. The diversity is directly related to the number (and respective probabilities) of different paths connecting a specific node to the rest of the network. Nodes with a few paths of fixed length connecting them to the remaining nodes have low diversity values, and tend to be placed at the "border" of the network. High diversity nodes have more balanced probabilities of being connected to other nodes through indirect paths of the same length, and tend to be away from the network border [20]. Following the above referenced works, we used the correlation between in-/out-degrees and activity (steady-state node occupation/visiting probabilities for the random-walk process) as the structure-dynamics relationship. As objects of study, we considered two directed networks, namely, the macaque cortex [9] (exhibiting moderate correlations between degree and activity) and a WWW map [19] (practically without any correlations between degree and activity). We were able to observe some remarkable increases in the correlation strengths due to the division of nodes into border and non-border groups. In other words, border effects were sufficient to explain the intricate correlations occurring in both networks studied here.

This paper is organized as follows. Section 2 contains all the supporting definitions for our experiments: first we introduce the network structure and dynamics (Sects. 2.1 and 2.2) and their interrelationships (Sect. 2.3), followed by a detailed definition of the network border (Sect. 2.4) and description of the networks studied (Sect. 2.5). Results and discussion are included in Sect. 3, while concluding remarks can be found in Sect. 4.

2   Materials and Methods

In this section, we define several different network features, including the straightforward degree and the more sophisticated diversity measurement. Also explained here is the network dynamics, which is represented by the simple random walk, further analyzed in conjunction with the node degree in order to evaluate the structure-dynamics interplay. Finally, this section also includes the description of some real-world networks adopted in our experiments.

2.1 Network Structure Basics

A network of N nodes and E directed edges is mathematically represented by the $N \times N$ adjacency matrix A whose elements $A(i,j) = 1$ indicate the presence of a directed link connecting node i to node j ($A(i,j) = 0$ if the nodes i and j are not connected or if $j = i$). The network is assumed to be connected, i.e. it consists of a single cluster of connected nodes. The most basic network structure indicator used in this work is the node degree, which is the number of edges connected to a given node. In a directed network, we distinguish between in-degree $k_{in}(i) = \sum_{j=1}^{N} A(j,i)$ and out-degree $k_{out}(i) = \sum_{j=1}^{N} A(i,j)$ of a particular node i. The average network in- and out-degrees are given by $\langle k_{in} \rangle = \langle k_{out} \rangle = E/N$. The length of the shortest path connecting node i to node j is denoted by $\ell(i,j)$. A path is a sequence of non-repeating nodes $p(n_1, n_L) = (n_1, n_2, \ldots, n_m, \ldots, n_L)$ such that $A(n_m, n_{m+1}) = 1$ ($m = 1, \ldots, L-1$), and the length of the path between nodes n_1 and n_L is $\ell(n_1, n_L) = L - 1$ (i.e. the number of edges in the path). The average shortest path length of a network is simply denoted by ℓ. A strongly connected component of a directed network is a subnetwork having all pairs of nodes (i,j) connected by at least one path in each direction (i.e. from i to j and from j to i). Other relevant network concepts and definitions can be found in the survey [6].

2.2 Random-Walk Dynamics

We have chosen the simple random walk on graphs to simulate the dynamics taking place on a network [13, 16, 3]. This model considers one agent randomly visiting nodes on a network following the Markovian probability $P(i,j) = A(i,j)/k_{out}(i)$ of leaving node i to visit node j (that is, the probability of taking the next move depends on the current state of the walker). The transition (stochastic) $N \times N$ matrix P completely describes the walking model. We are interested in the steady-state walking regime, where the number of visits per node reaches the equilibrium $\pi(i) = \lim_{t \to \infty} v(i)/t$, with $v(i)$ being the number of visits to node i after t walking steps. Notice that π is the eigenvector of P corresponding to eigenvalue 1: $\pi P = \pi$. For statistical normalization, we use the solution with $\sum_i \pi(i) = 1$, that is, $\pi(i)$ is the probability of finding the walker at node i in the steady-state regime, which is called here the *activity* of node i. Also worth pointing out is that P must be irreducible (i.e. the network has to be strongly connected) and aperiodic (i.e. the network must not be bipartite) for the solution of π to be unique [5].

2.3 Structure-Dynamics Interplay

The relationship between structure and dynamics is specifically represented here by the correlations between degrees (k_{in} and k_{out}) and activity (π). We aim to investigate how node connectivity influences network dynamics by addressing, e.g., a question if highly connected nodes coincide with those which are also highly active. These correlations are well-known in non-directed networks (where $A(i,j) = A(j,i)$), with the activity being directly proportional to the degree [10]: $\pi(i) = k(i)/\sum_{j=1}^{N} k(j)$ – notice that there is no distinction between in- and out-degrees. On the other hand, one sufficient condition for this perfect correlation to occur in directed networks is that, for every node i, $k_{\text{in}}(i) = k_{\text{out}}(i)$ [8]. Nevertheless, this is seldom the case in real-world directed networks [8] (see also Figs. 2 and 4). In order to quantify the strength of the (linear) structure-dynamics correlations, we use the Pearson correlation coefficient r [11], with $|r| \to 1$ for strong correlations and $|r| \to 0$ for weak correlations (the sign of r indicates a positive or a negative correlation).

2.4 Network Border

We now define the concept of diversity, which is associated with the definition of network border [20]. The diversity is based on the probability $P_h(i, j)$ of a walker going from node i to reach node j in h steps. This probability is computed by using self-avoiding random walks, where the walker is not allowed to return to nodes already visited (i.e. the walker follows only paths – defined in Sect. 2.1). We used a deterministic (exact) algorithm to compute matrices P_h instead of an approximated one [20]. The diversity (or diversity entropy) indicates how diverse the connectivity from (or to) a given node i to (or from) other nodes through paths of length h is. Considering that Ω is the set of all remaining nodes in the network (excluding i), the outward diversity of a node i, $D_{\text{out},h}(\Omega, i)$, is defined as follows:

$$D_{\text{out},h}(\Omega, i) = -\frac{1}{\log(N-1)} \sum_{\substack{j \neq i}}^{N} P_h(i, j) \log(P_h(i, j)) . \tag{1}$$

Conversely, the inward diversity takes into account all paths reaching node i:

$$D_{\text{in},h}(\Omega, i) = -\frac{1}{\log(N-1)} \sum_{\substack{j \neq i}}^{N} P_h(j, i) \log(P_h(j, i)) . \tag{2}$$

Notice that the diversity was previously defined for non-directed networks [20], where the distinction between inward and outward diversities is not necessary. In- and out-diversities are intuitively related to the concept of network border. Departing from marginal nodes in order to visit other nodes is not as effective as starting from non-border ones, since the former present fewer options in terms of possible paths (i.e. it is necessary to first leave the border through a limited number of paths in order to access the rest of the network). In this way, nodes not placed at the border tend to have higher diversity entropies than peripheral ones [20]. Therefore, the network border can be defined as the set of nodes characterized by diversity smaller than (or equal to) a certain network specific threshold value, T_E. Whether to use in- or out-diversities depends on the particular network under analysis.

2.5 Real-World Data

In the following paragraphs we briefly describe the networks used in this work, one biological and one technological. They present intricate correlations between degrees and activity, which we analyzed taking into consideration the border structure.

- *Macaque network*: Directed network of the macaque monkey brain considering the connectivity between cortical regions in one hemisphere [9]. Since we chose a simple random walk, the largest strongly connected component (see the end of Sect. 2.2) consisting of $N = 85$ nodes was used in our analysis. The respective average degrees are $\langle k_{in} \rangle = \langle k_{out} \rangle \simeq 27.72$ and the average shortest path length is $\ell \simeq 1.86$.
- *WWW network*: Directed network consisting of hyperlinks between webpages in the www.massey.ac.nz domain [19]. We again used the respective largest strongly connected component, with $N = 10810$ nodes, $\langle k_{in} \rangle = \langle k_{out} \rangle \simeq 14.63$ and $\ell \simeq 6.32$.

More details on the construction of these networks can be found in Refs. [9, 19].

3   Results and Discussion

In this section, we describe and discuss the results obtained by the analysis of the two networks mentioned in the previous section. The main approach adopted here is to investigate the correlations between structure and dynamics ($k_{in} \times \pi$ and $k_{out} \times \pi$ where symbol $\times$ means correlations) by computing the Pearson correlation coefficient either for all the nodes, border or non-border nodes. Interesting correlations were found between the in-degree k_{in} and the activity π (Figs. 1–4) – results for correlations between out-degree and activity were much less interesting (results not shown). Although these correlations are relatively weak when considering all network nodes (Pearson coefficient $r \simeq 0.66$ for macaque and $r \simeq 0.15$ for WWW), we observe that correlations tend to increase at the border/non-border regions. Before analyzing each network in details, we point out that the path probabilities P_h (see Sect. 2.4 for the border definition) were computed up to length $h = 4$ for both networks.

When using the out-diversity for border detection and varying the border threshold T_E in the range of the possible values of out-diversity, we found that correlations are higher inside and outside the border than in the entire macaque network – see all subplots in Fig. 1. Correlations inside the border are stronger than outside, with Pearson coefficients frequently around $r \simeq 0.85$ and $r \simeq 0.75$, respectively. Furthermore, analysis of the scatter-plot for node activity *vs.* in-degree (see Fig. 2 where we used out-diversity with $h = 4$ for identifying the border nodes) revealed a clear pattern of correlations for a specific border threshold: two separate correlations were found, one with a faster increase in activity for increasing in-degrees (inside the border, Pearson correlation strength $r \simeq 0.84$), and another with a slower activity increase (outside the border, $r \simeq 0.77$). Despite the presence of a couple of outliers, the difference between correlation slopes inside and outside the border is significant (very similar results were obtained for $h = 3$). Therefore, the intricacies observed in the correlation $k_{in} \times \pi$ in the macaque cortex can be explained, at least to some extent, by border effects.

A similar analysis was undertaken for the WWW network. In Fig. 3, we show correlation coefficients for $k_{in} \times \pi$ inside and outside the border against out-diversity thresholds. The strongest correlations inside the border were found for path length $h = 3$, with a remarkable increase to $r \simeq 0.9$. Strong correlations were also found outside the border for $h = 3$ and $h = 4$, with $r \simeq 1$. As for the macaque cortex, the WWW network shows

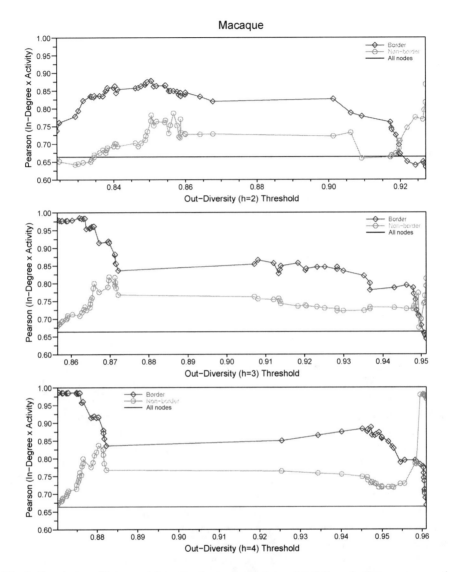

Fig. 1. Correlations (Pearson r) between the in-degree k_{in} and activity π in the macaque cortical network (y-axes) against the threshold T_E used for border detection with $h = 2$ (upper panel),
$h = 3$ (middle panel) and $h = 4$ (bottom panel). The values of T_E vary within the following ranges:
$0.824 \leq T_E \leq 0.927$ (top panel), $0.857 \leq T_E \leq 0.951$ (middle panel) and $0.870 \leq T_E \leq 0.961$ (bottom panel). Different curves indicate correlations inside and outside the network border, whereas
constant curves refer to the correlation coefficient $r \simeq 0.66$ for all network nodes. Data points are
not equally spaced on the x-axis because we have used only thresholds that occur in the respective
diversity distribution for the macaque network.

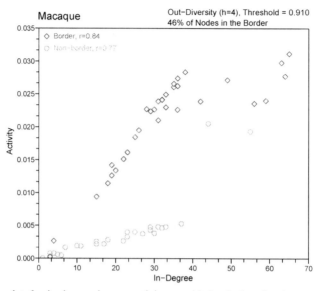

Fig. 2. Scatter-plot for in-degree k_{in} *vs.* activity π with border/non-border separation for the macaque cortical network, where different correlation tendencies were found for border and non-border nodes. The out-diversity was used for border detection (more details above the plot). The correlation coefficient for the entire network is $r \simeq 0.66$, while the correlation coefficients for border and non-border nodes are shown in the plot.

considerably stronger correlations at the border/non-border level than at the global network level. A detailed analysis of these correlations revealed that for some out-diversity thresholds (when $h = 4$) there are two distinct types of correlations for in-degree *vs.* activity: one for the non-border group, in which the nodes with very high in-degree (hubs of connectivity) are included, and another for the border group, where highly active nodes (called here as dynamical hubs) appear – see Fig. 4. Only the former correlations are strong ($r \simeq 0.85$), while the latter are remarkable for including dynamical hubs even though in-degrees in this group are relatively small. This specific WWW partition into border/non-border nodes is shown in Fig. 5, where it is possible to visually discriminate these two types of nodes. Border nodes are mainly placed inside (relatively) isolated groups at the periphery of the network, while other nodes are primarily placed at the well-connected core.

Notice that results using the in-diversity for identification of the border nodes were not as interesting as those showed in Figs. 1–4 for the out-diversity, with very low (or even absent) increases in the structure-dynamics correlations at the border/non-border level. Other distinctive behavior is that diversity values are higher for the macaque than for the WWW, which is possibly due to the very dissimilar relative densities of these networks: the average degree for the macaque is almost 28 (for only 85 nodes) while the average degree for the WWW is almost 15 (for thousands of nodes). Therefore, more paths are available between pairs of nodes in the macaque than in the WWW, leading to higher diversity entropies. To summarize, two distinct correlations between

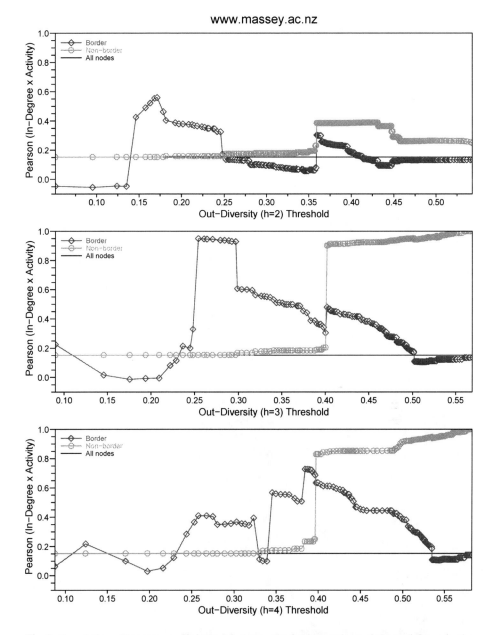

Fig. 3. Correlations (Pearson coefficient r) between the in-degree k_{in} and the activity π in the WWW network against threshold T_E used for border detection (see the caption to Fig. 1 for more information about the figure organization). Ranges of diversity threshold are $0.051 \le T_E \le 0.544$, $0.090 \le T_E \le 0.570$, and $0.087 \le T_E \le 0.583$, for $h = 2, 3, 4$, respectively. The correlation for all network nodes is $r \simeq 0.15$ (see constant curves). Data points are not equally spaced on the x-axis because we have used only thresholds that occur in the respective diversity distribution for the WWW network.

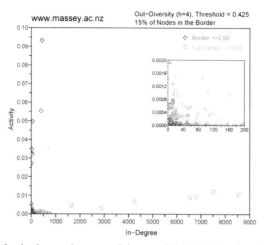

Fig. 4. Scatter-plot for in-degree k_{in} vs. activity π with border/non-border separation for the WWW network. This figure is organized in the same way as Fig. 2, with an additional inset magnifying an interesting part of the main graph. The correlation coefficient for the entire network is $r \simeq 0.15$.

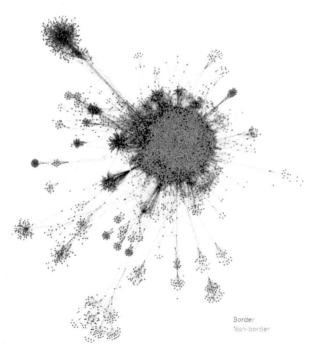

Fig. 5. Graphical representation of the WWW network generated by Cytoscape [17] using the force directed layout, where border and non-border nodes are distinguished by different colors. The out-diversity threshold used for border detection is $T_E = 0.425$ for path length $h = 4$ (the same as in Fig. 4). A similar picture for the macaque network exhibits a much denser network and the border/non-border boundary is not clearly detected only by visual inspection (not shown).

node activity and in-degree were found according to the division border/non-border (out-diversity based) for both networks (Figs. 2 and 4). The steeper increase in the node activity with increasing in-degree was observed for border nodes in the macaque cortex and also in the WWW, a result that may appear counterintuitive since highly active nodes are found to be almost exclusively at the border. One may argue that these nodes should appear outside the border, in the "center" of the network, where the random walker would possibly spend more time. However, our results are based on the out-diversity for border detection, i.e. only outgoing paths are considered for peripheral node analysis. In other words, inbound paths, which are the ones effectively providing activity to nodes, are not the ones defining borders. The number of outgoing paths for border nodes is less than for the central nodes by definition. This means that the number of routes to escape from the border nodes is less than for central ones and thus a random walker is more likely to be found on the border nodes. Further analyses, which are beyond the scope of this paper, could better elucidate the reasons for border nodes being the most active ones in the macaque cortex and in the WWW map here studied.

4   Concluding Remarks

We found that distinct and strong structure-dynamics correlations occur inside and outside the border in two networks, namely the macaque cortical network and the WWW map at the www.massey.ac.nz domain. Consequently, it was possible to better explain how structure and dynamics interact with each other in these networks: at the border, the node activity grows faster with increasing degrees than that for nodes outside the border. This effect is especially pronounced for the WWW network. With these results it is possible, depending on whether a node is inside or outside the border, to approximate the activity of a macaque cortical region (or a webpage) by only assessing its in-degree. The activity in the cortex can be thought as the amount of electric signals flowing onto each cortical region; in the WWW, it can be used to describe the level of user visits to each webpage. Other models of dynamics, perhaps more realistic for each type of network (e.g. a surfer model for the WWW), shall be investigated in the light of the border concept. Other further developments could help to understand these two coexisting and different dependencies of the node activity on in-degree. For instance, border and non-border properties can be compared employing other structural measurements, such as betweenness centrality and efficiency [6]. This approach could also shed light on the reasons for the highest active nodes being placed at the border. Another important issue is that when using the in-diversity for border detection, results are not so interesting as for the out-diversity. The specific features of the sample networks here employed leading to this behavior might be investigated in the future. Furthermore, other insights could be gained in experiments using classic network models (e.g. random, scale-free) or new models with controlled levels of border structure.

Acknowledgments. L. da F. Costa is grateful to FAPESP (05/00587-5) and CNPq (301303/06-1) for financial support. L. Antiqueira thanks the sponsorship of FAPESP (06/61743-7).

References

1. Arenas, A., Díaz-Guilera, A., Kurths, J., Moreno, Y., Zhou, C.: Synchronization in complex networks. Physics Reports 469(3), 93–153 (2008)
2. Barabási, A.L., Albert, R.: Emergence of scaling in random networks. Science 286(5439), 509–512 (1999)
3. Barrat, A., Barthélemy, M., Vespignani, A.: Dynamical processes on complex networks. Cambridge University Press, Cambridge (2008)
4. Boccaletti, S., Latora, V., Moreno, Y., Chavez, M., Hwang, D.U.: Complex networks: Structure and dynamics. Physics Reports 424(4-5), 175–308 (2006)
5. Bollobás, B.: Modern graph theory. Springer, New York (1998)
6. Costa, L.F., Rodrigues, F.A., Travieso, G., Villas Boas, P.R.: Characterization of complex networks: A survey of measurements. Advances in Physics 56(1), 167–242 (2007)
7. Costa, L.F., Sporns, O.: Correlating thalamocortical connectivity and activity. Applied Physics Letters 89(1), 013903 (2006)
8. Costa, L.F., Sporns, O., Antiqueira, L., Nunes, M.G.V., Oliveira Jr., O.N.: Correlations between structure and random walk dynamics in directed complex networks. Applied Physics Letters 91(5), 054107 (2007)
9. Kaiser, M., Hilgetag, C.C.: Nonoptimal component placement, but short processing paths, due to long-distance projections in neural systems. PLoS Computational Biology 2(7), e95 (2006)
10. Lovász, L.: Random walks on graphs: A survey. In: Miklós, D., Sós, V.T., Szőnyi, T. (eds.) Combinatorics. Paul Erdős is Eighty, vol. 2, pp. 353–398. János Bolyai Mathematical Society, Budapest (1996)
11. Myers, J.L., Well, A.D.: Research Design and Statistical Analysis. Lawrence Erlbaum Associates, New Jersey (2003)
12. Newman, M.E.J.: The structure and function of complex networks. SIAM Review 45(2), 167–256 (2003)
13. Noh, J.D., Rieger, H.: Random walks on complex networks. Physical Review Letters 92, 118701 (2004)
14. Pearson, J.E.: Complex patterns in a simple system. Science 261(5118), 189–192 (1993)
15. Rudnick, J., Gaspari, G.: Elements of the random walk: an introduction for advanced students and researchers. Cambridge University Press, Cambridge (2004)
16. Samukhin, A.N., Dorogovtsev, S.N., Mendes, J.F.F.: Laplacian spectra of, and random walks on, complex networks: are scale-free architectures really important? Physical Review E 77(3), 036115 (2008)
17. Shannon, P., Markiel, A., Ozier, O., Baliga, N.S., Wang, J.T., Ramage, D., Amin, N., Schwikowski, B., Ideker, T.: Cytoscape: A software environment for integrated models of biomolecular interaction networks. Genome Research 13(11), 2498–2504 (2003)
18. Tadić, B., Rodgers, G.J., Thurner, S.: Transport on complex networks: Flow, jamming and optimization. International Journal of Bifurcation and Chaos 17(7), 2363–2385 (2007)
19. Thelwall, M.: A free database of university web links: Data collection issues. Cybermetrics 6/7(1), 2 (2002)
20. Travençolo, B.A.N., Viana, M.P., Costa, L.F.: Border detection in complex networks. New Journal of Physics 11(6), 063019 (2009)
21. Watts, D.J., Strogatz, S.H.: Collective dynamics of 'small-world' networks. Nature 393(6684), 440–442 (1998)

Potential Link Suggestion in Scientific Collaboration Networks

Cristian K. dos Santos, Maurcio Onoda, Victor S. Bursztyn, Valeria M. Bastos, Marcello P.A. Fonseca, and Alexandre G. Evsukoff

Federal University of Rio de Janeiro COPPE/UFRJ
c.klen@coc.ufrj.br, monoda@uninet.com.br, victor@lb.com.br,
valeriab@ntt.ufrj.br, marcpa@centroin.com.br,
alexandre.evsukoff@coc.ufrj.br

Abstract. This works presents a methodology to suggest potential relationships among the elements in the scientific collaboration network. The proposed approach takes into account not only the structure of the relationships among the individuals that constitute the network, but also the content of the information flow propagated in it, modeled from the documents authored by those individuals. The methodology is applied it the accepted papers for the 2nd Workshop on Complex Networks - Complenet'2010. The results show insights on the relationships, both existent and potential, among elements in the network.

1 Introduction

Scientific collaboration networks is one of the most widely studied type of complex networks. Most of studies concern existing co-authorship networks, from which a number of structural properties have been derived [7][8]. The contents of the publications are not usually considered but they can provide information to suggest potential co-authorship in order to group together researches interested in the same subject.

The problem of potential link discovery in networks has been studied in the recent literature [1][6]. In the Relational Topic Model [1], the links are predicted using texts' content based on a probabilistic topic model [2], which allows inferring descriptions of the network's elements.

In this work, the co-authorship structure and the documents' content are used in order to analyze the network for link suggestion. The methodology is based upon the Newman [9] algorithm for community structure detection. The main contribution of this work is the methodology that integrates document clustering and community detection for the discovery of potential relationships in a co-authorship network. The methodology is applied to the accepted papers for the 2nd Workshop on Complex Networks – Complenet'2010 (most of them published in this volume).

The paper is organized as follows: next section presents the basic terminology of the community structure detection algorithm based on the recursive spectral optimization of the modularity function. Section three presents the main steps for representing a collection of documents as a documents network, where the nodes represent the documents and the links are weighted according to their similarity. Section four presents

L. da F. Costa et al. (Eds.): CompleNet 2010, CCIS 116, pp. 57–67, 2011.

the methodology itself of applying the community structure algorithm for potential link suggestion in co-authorship networks. Section five presents the results and section six the conclusions and final remarks.

2   Community Detection in Networks

The detection of community structure is one of the most active areas in the study of complex networks [10][11][13]. The concept of a good community partition is very difficult and can be formally defined in many different ways and many algorithms have been proposed. In the graph theoretic approach, the algorithms are usually formulated as a graph partition problem, in which the objective is to find the minimum cut in the graph. This approach is widely used for community structure detection in complex networks [9][12][13] and also is the base formalism for spectral clustering [14].

In the complex network literature, many methods to community structure detection employ the maximization of the function known as "modularity", introduced by Newman and Girvan [10]. This measure allows quantifying the "goodness" of possible subdivisions of a given network into communities and is the base for many algorithms [9][10][13]. The modularity measure is, however, not able to detect very small communities, as it has been recently pointed out by Fortunato and Barthlemy [5].

Community detection algorithms have been recently studied for document clustering [3][4], where the Newman algorithm [9] was compared to spectral clustering techniques. The first produced better results. In this section the main terminology is reviewed and the Newman algorithm for community detection is presented.

2.1   Modularity

Consider a graph $G(V,E)$, where $V = \{v_1, v_2, \ldots, v_n\}$ is the set of nodes and E is the set of edges connecting pairs of nodes. The graph's topology is represented by the adjacency matrix $A \in \{0,1\}^{nxn}$, whose elements are defined as:

$$a_{ij} = \begin{cases} 1 & if\{v_i, v_j\} \in E \\ 0 & otherwise \end{cases} , i = 1\ldots n, j = 1\ldots n \tag{1}$$

A community structure is a partition of the set V in c communities $C_1, C_2, \ldots, C_c$, such that $V = \bigcup_{i=1}^{c} C_i$ and $\bigcap_{i=1}^{c} C_i = \emptyset$. Newman and Girvan [10] defined the modularity as a quantitative measure to evaluate an assignment of nodes into communities. This measure can be used to compare different assignments of nodes into communities and is based on an association function $\phi(C_i, C_j)$ computed as:

$$\phi(C_i, C_j) = \sum_{i \in C_i} \sum_{j \in C_j} a_{ij} \tag{2}$$

The modularity function is then defined as [9]:

$$Q = \sum_{i=i}^{c} \left(\frac{\phi(C_i C_i)}{\phi(V,V)} - \left(\frac{\phi(C_i,V)}{\phi(V,V)} \right)^2 \right) \tag{3}$$

The term $\phi(C_i, C_i)$ measures the within-community sum of edge weights; $\phi(C_i, V)$ measures the sum of weights over all edges attached to nodes in community C_i and $\phi(V, V)$ is the normalization term that measures the sum over all edge weights in the entire network. Considering the adjacency matrix defined as (1), the term $\phi(C_i, C_i)/\phi(V, V)$ in the modularity function (3) is the empirical probability that both vertices of a randomly selected edge are in the subset (community) C_i. The second term $(\phi(C_i, V)/\phi(V, V))^2$ is the empirical probability that only one of the ends (either one) of a randomly selected edge is in the subset C_i. Thus, the modularity measures the deviation between observed cluster structure and the one that could be expected under a random model. If the number of within-community edges is no better than random, then the modularity is $Q = 0$. A modularity value $Q = 1$, which is the maximum, indicates strong community structure. In practice, typical values for actual networks are in the range from 0.3 to 0.7.

2.2 Spectral Modularity Optimization

There are several methods for finding community structure by maximization of the modularity function [9][10][12][15]. Nevertheless, the exact solution of the modularity optimization is argued to be a NP-hard problem, since it is formally equivalent to an instance of the MAX-CUT problem [9]. In general, an approximate solution must be employed. In this work, the spectral optimization approach proposed in Newman [9] is adopted as it has been shown good results in finding clusters in document networks, when compared with spectral clustering [3][4].

The spectral approach for modularity maximization is based on the definition of the modularity matrix $B \in R^{n \times n}$, computed as [9]:

$$B = A - P \tag{4}$$

Where A is the adjacency matrix (1) and P is the matrix that represents the random model, of which each element p_{ij} is the probability that the vertices v_i and v_j are linked in a random network, which is computed as:

$$p_{ij} = \frac{k_i k_j}{2M} \tag{5}$$

Where k_i and k_j are respectively the degrees of nodes v_i and v_j and $M = \frac{1}{2} \sum_i \sum_j a_{ij}$ is the total number of edges in the network.

The modularity function can thus be written as:

$$Q = \frac{1}{4M} \sum_{ij} b_{ij} \delta_{ij} \tag{6}$$

where b_{ij} is an element of matrix B and δ_{ij} is the Kronecker delta, such that $\delta_{ij} = 1$, if $i = j$ and $\delta_{ij} = 0$ otherwise.

In the case of partitioning the network in only two communities, the community structure can be represented by a vector $y \in \{-1, 1\}^n$, of which the elements are defined as:

$$Y_i = \begin{cases} 1 & if \ v_i \in C \\ -1 & if \ v_i \notin C \end{cases}, i = 1...n \tag{7}$$

The modularity can thus be written as:

$$Q = \frac{1}{4M} y^T B y \tag{8}$$

An approximate solution $u \in R^n$ for the modularity maximization problem can be computed by the maximization of the Rayleigh coefficient:

$$\hat{Q} = \frac{u^T B u}{u^T u} \tag{9}$$

The solution that maximizes (9) is the eigenvector corresponding to the largest eigenvalue of the matrix B, which is computed as the solution of the eigenvalue problem $Bu = \lambda u$. An approximation of the indicator vector is then obtained by the sign of the eigenvector component:

$$\hat{y}_i = \begin{cases} 1 & if \ u_i \geq 0 \\ -1 & if \ u_i > 0 \end{cases}, i = 1...n \tag{10}$$

A general network, however, contains more than two communities. This approach can thus be applied recursively to find a partition of the network into more than two communities. The idea of the recursive algorithm is to evaluate the gain in the modularity function if a community is further divided. For any community C' generated by a partition (10), the additional contribution to the modularity ΔQ is computed as:

$$\Delta Q = \frac{1}{4M} y'^T B' y' \tag{11}$$

where y' is the partition vector that will further subdivide the community C'. The modularity matrix B' is the sub-matrix of B considering only the vertices that belong to C'. The elements b'_{ij} of the matrix B' are computed at each recursive step as:

$$b'_{ij} = b_{ij}\delta_{ij} - \sum_{k \in C'} b_{ik} \tag{12}$$

Where b_{ij} are the elements of the modularity matrix computed for the community to be divided and δ_{ij} is the Kronecker delta. The indexes (i, j) refer to the nodes of the entire network, i.e. $i,j = 1 \ldots n$.

The recursive process is halted if there is no further division of a community in the network that will increase the total modularity function, and therefore there is no gain in continuing to divide the network. In practice, the test $\Delta Q < 10^{-3}$ is used as the stopping criterion. This approach is interesting because it does not depend on the number of communities as input, as this is obtained as an output of the algorithm.

The recursive community detection algorithm is sketched bellow. The algorithm is initialized with the modularity value $Q = 0$ and number of clusters $c = 0$. The algorithm executes itself recursively to compute the indicator matrix $Y \in \{-1,1\}^{n \times c}$ that stores the result of the partition computed as (10). The outputs of the algorithm are the final number of communities c and the final value of the modularity function Q.

The community detection algorithm is used to find clusters in documents networks as presented in next section. The clusters of documents are then used to identify potential co-authorship is scientific collaboration networks, discussed in section four.

Algorithm 1. The recursive spectral community detection		
Input:	The matrix A, the modularity value Q , the number of communities c	
Output:	Community indicator matrix Y; the number of communities c and the final value of the modularity function Q.	
01	**Begin**	
02	Compute B for C_0 as (4)	
03	Compute the eigenvalue problem $Bu = \lambda u$;	
04	Compute y as (10) to split C_0 in two communities C_1 and C_2;	
05	Compute ΔQ for B and y, according to (8);	
06	**If** $\Delta Q > 0$	
07	Define $Q \leftarrow Q + \Delta Q$;	
08	Apply recursively the algorithm to C_1, Q and c;	
09	Apply recursively the algorithm to C_2, Q and c;	
10	**Else**	
11	Define $c \leftarrow c + 1$;	
12	Update the matrix Y with y at the position of c;	
13	**End if**	
14	Return Y, c and Q	
15	**End**	

3   Document Networks

Document clustering is one of the most active research application areas in text mining. Document clustering involves two phases: first, the feature extraction maps each document to a point in high-dimensional space and then, clustering algorithms group the points into a clustering structure. The feature extraction phase often uses the vector-space representation of the document collection where the unstructured content is structured into a numeric table, whose lines are related to the documents and columns are related to the words that appear in the entire collection. The document collection is usually preprocessed with the elimination of irrelevant words ("stop words") and the application of stemming algorithms to reduce similar words to one semantically equivalent term. The collection of documents is thus represented by the set $D = \{d_1, d_2,\ldots,d_n\}$ and the set of relevant terms is denoted as the set $T = \{t_1, t_2,\ldots,t_m\}$.

In the vector-space representation, a document $d_i \in D$ is represented as a vector $x_i \in R^m$, of which an element x_{ij} represents the importance (usually the frequency) of the term t_j in the document d_i. The entire document collection is thus represented as a matrix $X \in R^{nxm}$, sometimes referred as the Bag-of-Words (BoW).

The elements of the BoW are computed by the tf.idf as:

$$x_{ij} = \frac{f_{ij}}{\sum_{k=1}^{m} f_{ij}} log \left(\frac{n}{n_j} \right), i = 1...n, j = 1...m \qquad (13)$$

Where f_{ij} is the number of occurrences of the term t_j in the document d_i and n_j is the number of documents where the term t_j has appeared at least once (or other pre-defined value).

In a document network, each node $v_i \in V$ represents a document and an edge is weighted by the similarity between two documents, computed by their vector-space representation. The document clustering task corresponds thus to the community detection in the document network.

In order to apply the community detection algorithm presented above for document clustering it is necessary to reduce the weighted document network into an un-weighted graph $G(V, \varepsilon)$ by defining the adjacency matrix elements $a_{ij} \in A$ as:

$$a_{ij} = \begin{cases} 1 & \text{if } i \neq j \text{ and } h(x_i, x_j) \geq \xi \\ 0 & \text{otherwise} \end{cases} \tag{14}$$

where ξ is a threshold value that determines what should be considered as similar documents. The threshold ξ is very important since different threshold values result in different topologies and thus different community structures for the same document collection. The effect of the threshold in the present case study is discussed in section 5. The function $h(x_i, x_j)$ measures the similarity between the documents represented by the vectors x_i and x_j and, as usual in text mining literature, the cosine metric is applied:

$$h(x_i, x_j) = \frac{\langle x_i, x_j \rangle}{\sqrt{\langle x_i, x_i \rangle \langle x_j, x_j \rangle}} \tag{15}$$

where $<x_i, x_j> = x_i^T x_j$ is the dot product.

The community detection algorithm is applied in the documents and authorship networks in order to find the potential co-authorship relations based on the documents collection, as described in next section.

4 Potential Link Discovery in Co-authorship Networks

In this work there are three sets of objects: the set of individuals; the set of documents, of which each document is related to one or more authors, and the set of terms appearing in the documents. Three kinds of networks can be defined from theses objects' sets. The first network represents the existing relationship among individuals. In the case presented in this work, individuals are the authors and the links in the existing relationship network represent the co-authorship between two authors in at least one paper presented in the Workshop. In a more general setting, the existing relationships can be represented by other kind of networks, such as hyperlinks in a set of blogs, friendship in a social network web service, emails, etc.

The second network is the document network, in which the nodes represent the documents and the edges represent similar documents, as determined by the terms appearing in the documents, as discussed in the previous section.

The third network is the potential relationship network, which is a combination of the previous ones. In the potential relationships network, the nodes are the individuals and each link is weighted by the similarity between the content produced by the two individuals. In the vector space representation, the weight of a term for each author is computed as the sum of the tf.idf values of all documents produced by the author as:

$$z_k = \sum_{i \in D_k} \frac{x_i}{N_i}, \quad k = 1...N \tag{16}$$

where $D_k \subset D$ is the subset of documents authored by each one of the $k = 1 \ldots N$ authors. An author can participate on more than one document, such that the tf.idf value for each document must be divided by the number of co-authors of that document N_i.

The potential relationship network is then obtained as before (eq. (14)) for document networks, where the cosine similarity (15) is used to define the links corresponding to the vector-space representation of authors as (16). Finally, the edges in the existing relationship network must be subtracted from the potential relationship network, such that all the links in this network do not yet exist.

The suggestion of potential links must be done for individuals that are related to each other. The community detection algorithm is applied to the potential relationship network such that each community represents related content produced by the authors within. Thus, two authors in the same community indicate that they are interested in the same subject and a link could be suggested between them.

The proposed approach is sketched as follows. The document collection provides information for the co-authorship network and the document network. The combination of both of these networks allows the definition of the potential collaboration network, as discussed above. The potential links are suggested from the community detection on the collaboration network by identifying authors that have published documents on the same subject but are not yet co-authors. This approach is further discussed for the case of study and presented in the next section.

5 Results

The method for potential link suggestion presented above has been applied to abstracts of the papers accepted for the 2nd Workshop of Complex Networks, most of them are in this volume. This section presents the results of the two main steps of the methodology: document clustering and potential link suggestion.

5.1 Document Clustering

The document network showing the clustering structures computed and the algorithm is shown in Fig. 1, where the nodes represent the documents (accepted papers to the Workshop) and the links are similarities defined as (14) using a threshold value $\xi = 0.20$. The community detection algorithm has resulted in five clusters. The legend in Fig. 1 shows the percentage of documents in each group and the most relevant words in the tag cloud, where the size of a word is proportional to its importance in the group.

The groups found by the algorithm represent the main subjects explored by the papers in the Workshop. The visualization algorithm used in Fig. 1 places the nodes of the document network according to the edges' weights, i.e. the similarities among documents. The first cluster is related to community structure algorithms and applications and has resulted in a highly connected cluster, showing that the documents share many terms. The other clusters are sparser since a variety of subjects, from biological networks to trade and information networks have been exploited in the Workshop.

5.2   Potential Links

Two potential collaboration networks are shown in Fig. 2 and Fig. 3, where nodes represent the authors and the edges represent similarities between their papers defined as (14) using respectively $\xi = 0.25$ and $\xi = 0.30$ as thresholds. Two types of links are highlighted in the figure besides the similarity edges in gray. The links highlighted in blue represent the existing co-authorship relationships in at least one paper. The links highlighted in red are the potential co-authorship suggestions, i.e. authors that are in the same community but do not have co-authored a paper.

The community structure algorithm has found seven communities for the network defined with $\xi = 0.25$ (shown in Fig. 2) and ten clusters for the network defined with $\xi = 0.30$ (shown in Fig. 3). Usually, the greater the threshold, the smaller is the number of edges and the greater is the number of communities found by the algorithm. nevertheless this is not straightforward and depends on the problem.

The high interconnectivity of the group related to "community structure algorithms" found in the document network (Fig. 1) also appears in the collaboration networks as the group G7 in the seven groups' network (Fig. 2) and the group G10 in the ten groups'

Fig. 1. The community structure of the document networks

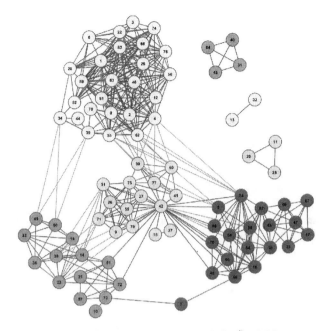

Fig. 2. The potential links network for $\xi = 0.25$

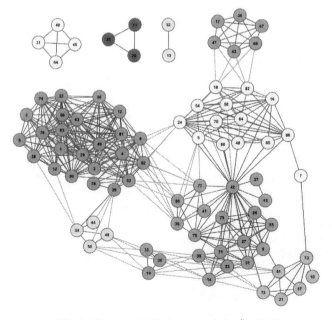

Fig. 3. The potential links network for $\xi = 0.30$

network. For these groups, the number of suggested links is greater since the methodology suggests a link to most of the authors within the same community as their papers are highly similar.

6   Conclusions

This work has presented an approach to identifying potential collaborations in co-authorship network. The Newton algorithm avoids the pre-definition of the number of clusters, which is a recurrent problem in cluster analysis. Moreover, the method is recursive so that it can deal with hierarchical structure, a feature frequently found in document clustering problems.

The method was applied to the abstracts of the papers accepted for the 2nd Workshop of International Networks, whose majority of papers has been published in this volume. The results obtained with the proposed method have guided to a better understanding of the main research subjects discussed in the workshop and also have revealed potential relationships among authors.

This research project will continue in the direction of scalability of the present analysis so that it can be applied in a wider network.

Acknowledgements. The financial support for this research has been provided by the Brazilian Research Agencies, CNPq, CAPES and FINEP.

References

1. Chang, J., Blei, D.M.: Relational topic models for document networks. In: AISTATS 2009, Clearwater Beach, Florida, USA (2009)
2. Chang, J., Boyd-Graber, J., et al.: Connections between the lines: augmenting social networks with text. In: KDD 2009, Paris, France (2009)
3. dos Santos, C.K., Evsukoff, A.G., de Lima, B.S.L.P.: Cluster analysis in document networks. WIT Transactions on Information and Communication Technologies 40, 95–104 (2008)
4. dos Santos, C.K., Evsukoff, A.G., de Lima, B.S.L.P.: Spectral clustering and community detection in document networks. WIT Transactions on Information and Communication Technologies 42, 41–50 (2009)
5. Fortunato, S., Barthélemy, M.: Resolution limit in community detection. Proceedings of the National Academy of Sciences 104(1), 36–41 (2007)
6. Liben-Nowell, D., Kleinberg, J.: The link prediction problem for social networks. In: CIKM 2003. ACM, New Orleans (2003)
7. Newman, M.E.J.: Scientific collaboration networks: I. Network construction and fundamental results. Phys. Rev. E 64, 016131 (2001)
8. Newman, M.E.J.: The structure of scientific collaboration networks. Proc. Natl. Acad. Sci. USA 98, 404–409 (2001a)
9. Newman, M.E.J.: Modularity and community structure in networks. PNAS 103(23), 8577–8582 (2006)
10. Newman, M.E.J., Girvan, M.: Finding and evaluating community structure in networks. Physical Review E 69(2), 26113–26115 (2004)
11. Ruan, J., Zhang, W.: Identifying network communities with a high resolution. Physical Review E 77(1), 16104–16112 (2008)

12. Wang, G., Shen, Y., et al.: A vector partitioning approach to detecting community structure in complex networks. Comput. Math. Appl. 55(12), 2746–2752 (2008)
13. White, S., Smyth, P.: A spectral clustering approach to finding communities in graphs. In: SIAM International Conference on Data Mining, pp. 76–84 (2005)
14. Xiang, T., Gong, S.: Spectral clustering with eigenvector selection. Pattern Recognition 41(3), 1012–1029 (2008)
15. Zarei, M., Samani, K.A.: Eigenvectors of network complement reveal community structure more accurately. Physica A 388(8), 1721–1730 (2009)

Deciding on the Type of a Graph from a BFS

Xiaomin Wang, Matthieu Latapy, and Michèle Soria

LIP6-CNRS-UPMC, 4, Place Jussieu, Paris, France
wangix@msn.com, {matthieu.latapy,michele.soria}@lip6.fr

Abstract. The degree distribution of the Internet topology is considered as one of its main properties. However, it is only known through a measurement procedure which gives a biased estimate. This measurement may in first approximation be modeled by a BFS (Breadth-First Search) tree. We explore here our ability to infer the type (Poisson or power-law) of the degree distribution from such a limited knowledge. We design procedures which estimate the degree distribution of a graph from a BFS of it, and show experimentally (on models and real-world data) that this approach succeeds in making the difference between Poisson and power-law graphs.

1 Introduction

The Internet topology is at the core of an intense research activity. Its *degree distribution*, *i.e.* the fraction P_k of nodes with k links, for all k, is of high interest: it may have a strong influence on the robustness of the network, on protocol design, and on spreading of information. Moreover, it is often claimed that it may deviate significantly from what classical models assume, which leads to an intense activity on modeling issues.

However, the degree distribution of the Internet topology is not readily available: one only has access to *samples* of this graph, obtained through measurement procedures which are intricate, time and resource consuming, and far from complete. Even more important, these samples are *biased* by the measurement procedure, which may have a strong influence on the observed degree distribution [8,5,2].

We explore here the following approach: we consider a simple model of Internet topology measurements and try to derive the type of the degree distribution of the underlying graph from the observation. Our basic goal therefore is to answer the following question: given limited information obtained from measurement, does the underlying topology more likely have a power-law or a Poisson degree distribution? In many cases, the measurement process may be approximated by a BFS (Breadth First Search) tree from a given node of the network. In the lack of a widely accepted and better solution, this approximation has been used in many occasions [2,5,6,8]. Deciding on the degree distribution from such a view is a challenging task. Here we first assume that the *size* of the graph, *i.e.* its number of nodes n, is given (a reasonable assumption [6]). In addition, we assume that the underlying graph is a random graph with either a Poisson or power-law degree distribution. And finally, we assume that we have a *complete* BFS of the considered graph. It must be clear that these two last assumptions are very strong, and are not attainable in practice. We however consider them as reasonable for a first approximation, and give hints of how to get rid of them in the last section of the paper.

L. da F. Costa et al. (Eds.): CompleNet 2010, CCIS 116, pp. 68–74, 2011.

2 Methodology

We aim at deciding the type of the degree distribution of an unknown graph G from one of its BFS; to do so, we build from the BFS a set of graphs according to different *parameters* of the expected degree distribution. Our expectation is that, if we use during this process parameters close to the ones of the original graph G, the degree distribution of the obtained graph should be close to the one of G. However, it is hard to find a general strategy for all types of graphs. Here we focus on *two typical types*, Poisson and power-law. Starting with a BFS T of size n and parameters of a degree distribution (Poisson with λ or power-law with α), the central idea is to iteratively add $m - n + 1$ edges to T in order to build a graph G' in such a way that, if the original graph is of the supposed type then G' will have a degree distribution similar to the original one. We define the strategies for doing so in next section.

We experimentally assess the validity of the approach by applying it to cases where we know the original graph G (Section 4 and 5). We then compare the degree distributions of G and the various G' obtained from each strategy and check conformance of results with expectations. To compare distributions, we use graphic plots of the inverse cumulative distributions, together with the classical Kolmogorov-Simirnov statistical test.

In Section 6, we extend our approach to get information on the degree distribution of a graph when the number of edges is not known, which is a more realistic assumption. In that case we use our reconstructing strategies for a wide range of possible values of m and infer the most probable type of degree distribution.

3 Rebuilding Strategies

We propose a method which, given a BFS and some information on a graph G, builds a graph G' *similar* to G concerning the degree distribution. We first assume that apart from the BFS, we know the number of vertices n, the number of edges m and that G has either a Poisson or a power-law degree distribution. Note that the BFS T of G (for a connected graph) contains all n vertices, and $n - 1$ edges. Therefore, our reconstruction procedures will consist in adding $m - n + 1$ edges to T, with a strategy for choosing these links which depends on the supposed type of G.

Poisson graphs (RR strategy). Suppose G is a Poisson random graph. It may therefore be seen as the result of an ER (Erdös Rényi) construction [3]: starting with n vertices and no edge, m edges are uniformly chosen among the $\frac{n(n-1)}{2}$ possible pairs of vertices. The expected degree distribution obtained follows a Poisson law: $P_k = \frac{\lambda^k e^{-\lambda}}{k!}$ where $\lambda = \frac{2m}{n}$ is the average degree. We may think of building a graph G' similar to G by using a variant of the ER construction: starting with the n vertices and $n - 1$ edges of T, the $m - n + 1$ missing edges are randomly added as in the ER model. In this approach, though, T is a BFS of G but may not be a possible BFS of G': any edge in E which is not in T necessarily is between two vertices in consecutive levels[1] of T, or in the same level of T (otherwise T would not be a shortest path tree and thus not a

---

[1] The j-th level of a tree is the set of vertices which are at distance j from its root. Level j and $j + 1$ are said to be *consecutive* for any j.

BFS). In order to ensure that T is also a BFS of G' we will therefore add edges only between vertices in consecutive or in the same level. Since both extremities of edges are randomly chosen, we call this construction the RR (Random-Random) strategy. It is notable that the expected degree distribution obtained with this construction can be explicitly determined, so that one may estimate the expected degree distribution of G' from n, m and T, without explicitly constructing G'.

Power-law graphs (PP and RP strategies). Suppose now that G is a power-law graph. We therefore aim at designing a process which will build, from a BFS T, a graph G' with power-law node degree distribution. To do this, we add $m - n + 1$ edges between nodes in appropriate levels of T. However, these pairs of nodes are no longer chosen uniformly at random. Instead, we use a selection scheme inspired from the preferential attachment of the classical BA (Barabási-Albert) model [1]: we choose nodes randomly with a probability proportional to their degree in T. As we choose both extremities of added edges according to preferential attachment, we call this procedure the PP strategies. Finally, we also consider the RP (Random-Preferential) strategy in which we choose (in the appropriate levels) one vertex uniformly at random and the other one according to preferential attachment. This strategy has the advantage to lead to degree distribution which are not pure Poisson nor power-law distributions, but a mixture of both, which is likely to appear in practice. For both these strategies, an explicit computation of the expected degree distribution can also be done.

4   Validation Using Model Graphs

Our expectation is that the strategies described in the previous section succeed in building a graph G' similar (on degree distribution) to G when the appropriate strategy is used with an appropriate graph (RR if G is Poisson, PP if G is power-law). In addition, we expect that the degree distribution of G' will differ significantly from that of G if a wrong strategy is applied (RR when G is power-law, PP if G is Poisson). We experiment this hypothesis on *model graphs, i.e. random graphs* in the classes of Poisson graphs or power-law graphs with given parameters. To ensure the BFS will cover all nodes of the graph, we use a program which generates *random simple connected graphs* according to some degree sequence [7]. For each random graph G, we first extract a BFS, then build G'_{RR}, G'_{RP} and G'_{PP}, and compare them to G with respect to their degree distribution. In order to compare these distributions, we first plot *Inverse Cumulative Degree Distributions* $(ICDD(k) = \sum_{i \geq k} P_i)$, which are much more readable than the degree distributions themselves. We also compute a statistical test aimed at comparing distributions: the KS (Kolmogorov-Simirnov) test, defined as $KS = \max_j \left| \sum_{i=1}^{j} (P_i - Q_i) \right|$ for two distributions P_k and Q_k. Lower value of KS means less difference between the two degree distributions.

We conducted experiments with Poisson model graphs with average degrees from 3 to 10, and with power-law model graphs with exponents from 2.1 to 2.5. In each case, we consider graphs with 10000 and 100000 vertices, and all results in tables and figures are averaged over ten samples.

Poisson graphs. We present results for Poisson graphs with average degree 10 in Figure 1. The degree distribution obtained with RR strategy seems closer to the original one, as expected. This is confirmed by the KS statistics (Table 1): the smallest values are obtained with RR strategy. A Poisson graph with a higher degree gives better results (the ratio of RR to PP or RP). This is probably due to the fact that we add many more edges in this case, and so strategies for doing this make much more difference.

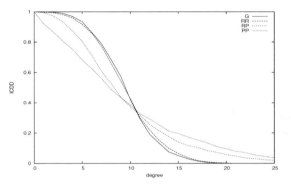

Fig. 1. Reconstruction of a Poisson 10. We draw the ICDD for four graphs: G, G'_{RR}, G'_{RP}, G'_{PP}, which show RR is the best reconstruction strategy for Poisson case.

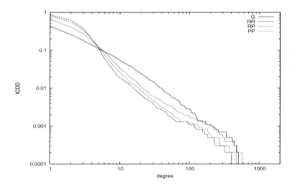

Fig. 2. Reconstruction of a power-law 2.2. We draw the ICDD for four graphs: G, G'_{RR}, G'_{RP}, G'_{PP}, which show PP is the best reconstruction strategy for power-law case.

Finally, we conclude that our reconstruction strategies succeed in recognizing random Poisson graphs. This is true for all average degrees, but performs best on graphs with a relatively high average degree.

Power-law graphs. We present the results obtained for power-law graphs with a typical exponent $\alpha = 2.2$, in Table 1 and Figure 2. They lead to graphs with expected average degree 3.14, but with huge variance. Again, all these results are in accordance with expectations for all tested exponents : PP strategy performs better than others. This shows that, as for Poisson random graphs, our reconstruction strategies succeed in recognizing power-law random graphs.

Table 1. KS for Poisson random graphs and power-law random graphs

	Poisson 3		Poisson 10		Power-law 2.1		Power-law 2.2	
n	10000	100000	10000	100000	10000	100000	10000	100000
RR	0.0078	0.0167	0.0162	0.0290	0.58560	0.46555	0.48682	0.38913
PP	0.0683	0.1159	0.2624	0.2530	0.25510	0.30156	0.26714	0.29962
RP	0.0470	0.0351	0.1652	0.1800	0.50095	0.58778	0.43562	0.46625

5   Validation Using Real-World Data (Skitter Graph)

In practice, considered graphs have neither perfect Poisson nor power-law degree distribution, and are not random. We consider have a real-world dataset among the current largest available measurements of the Internet using traceroute-like probes, the Skitter project of CAIDA [4]. Although the obtained graph still is a partial view and probably strongly biased, it constitutes current state-of-the-art of data available on Internet topology and we use it as a benchmark here, with 1719307 nodes and 12814336 links.

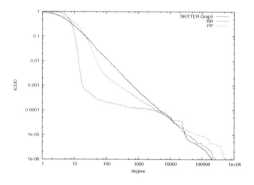

Fig. 3. ICDD for Skitter experiments: original graph, RR and PP strategies

Table 2. KS for Skitter experiments

	KS
Skitter-RR	0.28472278
Skitter-PP	0.11426863

Figure 3 shows the ICDD of our strategies (here we give only RR and PP, because of computation costs on data of such a large size). PP behaves best, which means that the entire Skitter graph follows a degree distribution of type power-law. This is confirmed by the values of KS in Table 2.

6   Deciding the Type from n and BFS

The goal of this section is to decide the type of degree distribution of an unknown graph G, using n and one of its BFS T only (no m). We proceed as follows: for each

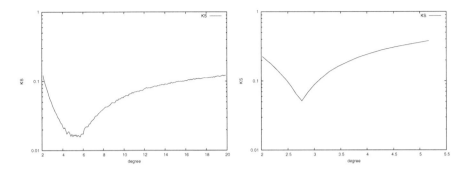

Fig. 4. KS difference for a random Poisson graph with average degree 3 reconstructed by the RR strategy (left) and by the PP strategy (right). The smallest value of KS is obtained for the RR strategy (0.0163 in RR and 0.0515 in PP), and it corresponds to an average degree 5.2.

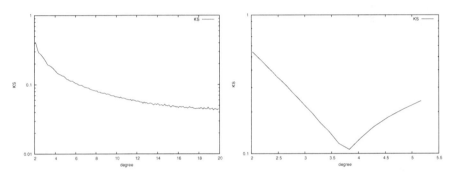

Fig. 5. KS difference for a random power-law graph with exponent 2.2 reconstructed by the RR strategy (left) and by the PP strategy (right). With the RR strategy, there is no minimum in our test range from 2 to 20. With the PP strategy, the lowest KS is at average degree 3.83.

reconstruction strategy, we compute the KS statistics for a wide range of realistic values of m. We then plot these statistics as a function of m and select the value of m which gives the best result (lowest KS). We then compare the results for different strategies and conclude on the most probable type of G. Figure 4 displays the results for an initial Poisson random graph G with average degree 3. Both RR and PP strategies identify a most relevant value for m (corresponding to average degree 5.2 for RR and exponent 2.26 or average degree 2.77 for PP). Moreover, the RR strategy leads to lower values of the KS test, which finally leads to the conclusion that G looks like a Poisson graph with average degree close to 5.2. This is satisfactory as indeed G is Poisson, but the estimate of the average degree is poor. A possible explanation of this phenomenon is that the power-law degree distribution of BFS (as shown in [5]) may lead to overestimate the value of the average degree.

The results when G is a power-law random graph are quite different. We plot a typical case (exponent 2.2, average degree 3.14) in Figure 5. We observe that the PP strategy succeeds in identifying a most relevant value for m (corresponding to exponent 2.12 or average degree 3.83 here), while the RR strategy leads to a continuous decrease of the KS statistics when m grows. As in Section 3, this is due to the fact that we add many links during the reconstruction. This leads to a strong impact of the strategy over the initial information (n and T). It is therefore important to keep the value of m reasonably small, and our strategies perform well only for such m.

In all the cases, though, the shapes of the plots obtained with RR strategies are very different on Poisson and power-law graphs. This may be used to distinguish between Poisson and power-law graphs, which was our goal: if RR strategy leads to a continuously decreasing plot then G is power-law, else it is Poisson.

7   Conclusion

We presented a new approach to derive the type of the degree distribution of a graph when only a BFS and its size are known. We use different strategies to reconstruct a graph from a BFS, according to the presupposed type of the graph. We then show how these strategies allow to decide between Poisson and power-law degree distributed graphs, with experiments on random graphs as well as real-world graphs. The main limitation of our contribution is that we suppose the knowledge of a complete BFS of the unknown graph. This hypothesis is not realistic, since practical measurements rather provide only paths to a subset of the real graph vertices. Our main perspective therefore is to reduce requirements on data, and design strategies needing truncated BFS (BFS until a certain level) only, n and partial BFS (that do not contain all vertices of the graph). Assuming that such measurements are available is much more realistic [9].

In future work, we also want to improve our work by investigating various refinements of the approach: (i) design reconstruction strategies taking into account more subtle properties of the BFS, such as node distribution on each level, and also other local properties such as the clustering coefficient; (ii) extend the reconstruction strategies in the direction of other types of distributions: a first step was made with the RP strategies for mixed Poisson-power-law graphs, but we also aim at investigating power-laws with exponential cut-off, and other laws.

References

1. Barabasi, A.-L., Albert, R., Jeong, H.: Mean-field theory for scale-free random networks. Physica A: Statistical Mechanics and its Applications (1999)
2. Lakhina, A., Byers, J.W., Crovella, M., Xie, P.: Sampling biases in ip topology measurements. In: INFOCOM 2003 (2003)
3. Bollobas, B.: Random Graphs. Cambridge University Press, Cambridge (2001)
4. CAIDA. skitter project
5. Achliotas, D., Clauset, A., Kempe, D., Moore, C.: On the bias of traceroute sampling: or, power-law degree distributions in regular graphs. Journal of the Association for Computing Machinery (2005)
6. Viger, F., Barrat, A., Dallasta, C.L., Kolaczyk, E.D.: What is the real size of a sampled network? the case of the internet. Physical Review E 75, 056111 (2007)
7. Viger, F., Latapy, M.: Random generation of large connected simple graphs with prescribed degree distribution. In: COCOON 2005 (2005)
8. Dallasta, L., A-hamelin, I., Barrat, A., Vazquez, A., Vespignani, A.: Exploring networks with traceroute-like probes: theory and simulations. In: TCS (2004)
9. Latapy, M., Magnien, C., Ouedraogo, F.: A radar for the internet. In: Proceedings of ADN 2008 (2008)

Generating Power Law Distribution of Spatial Events with Multi-agents

Vasco Furtado and Douglas Oliveira

Universidade de Fortaleza, Mestrado em Informática Aplicada,
Av. Washington Soares, 1321, Fortaleza, Brasil
vasco@unifor.br, dougpetcomp@yahoo.com.br

Abstract. The automated generation of events that follows a Power Law (PL) distribution has been extensively researched in order to mimic real world phenomena. Typically, the methods pursuing this goal consist of a unique generating function able to reproduce the inner features of PL distributions. On the contrary, most events that follow a PL distribution are produced through the interaction of different and distributed agents, which are often independent, autonomous, and have a partial perception of the world that surrounds them. In this paper, we investigate the circumstances in which multi-agents, in particular their communication mechanisms, produce spatial events that follow a PL. We are going to focus on models in which the agent's behavior is based on the ant colony optimization algorithm. We show that restricted models of agent communication based exclusively on pheromone exchange require an extension to represent direct communication in order to generate PL data distributions.

Keywords: Multi-agent Systems, Ant Colony Optimization, Power Law.

1 Introduction

Modeling the interconnected nature of social, biological and communication systems as complex networks or graphs has attracted much attention in diverse domains such as computer science [4], statistical physics [2] and biology [3]. A common property of many large networks is that the connectivity of vertices follows a Power Law (PL) distribution. As demonstrated by [4], such a feature is due to the fact that new vertices attach preferentially to already well-connected ones. Moreover, these studies have shown that networks continually expand through the addition of new vertices and links. The aforementioned preferential attachment hypothesis states that the rate $\Pi(k)$ with which a node with k links acquires new links is a monotonically increasing function of k and that $\Pi(k)$ follows a Power Law. Growth and preferential attachment are considered sufficient to explain the distribution pattern that governs this self-organizing phenomenon. Actually, this kind of distribution is verified not only in complex networks but also in several natural and artificial phenomena.

Based on these notions, methods for generating PL distributions in order to mimic real evolution of events have been proposed recently [5]. Such methods basically consist of a unique generating function able to reproduce the inner features of PL

L. da F. Costa et al. (Eds.): CompleNet 2010, CCIS 116, pp. 75–84, 2011.

distributions. In this paper, we intend to go further in the analysis of methods for generating PL distributions. We argue that typically, in real life, the generation of a global Power Law distribution for a certain phenomenon (we are going to dub it here an event) is not a product of the application of a unique function executed by a centralized agent. On the contrary, most events that follow a PL distribution are produced through the interaction of several agents which are often independent, autonomous, and have a partial perception of the world that surrounds them. This is particularly salient in social phenomena where agents are limited in terms of perception to the space they live in, and so the self-organizing effect (i.e. distribution of spatial events following a PL) is obtained from individual contributions.

Agent-based simulation has been shown to be an adequate tool for the modeling and analysis of the emergence phenomenon [11]. In this paper, we investigate the circumstances in which multi-agents, in particular their inter-agent communication mechanisms, produce spatial events that follow a PL. Our claim is that agents have the capability to pursue self-organized behavior by considering their individual (local) activities as well as the influence of others in the community they live in. The research questions that emerge from this investigation are: Is it possible to simulate a PL distribution of spatial events only with preferential attachment modeled individually in each agent? What is the role of communication of local-constrained agents for producing a global PL distribution of events? Is there any particular communication network topology that favors the generation of PL? In particular, we are going to focus on models in which the agent's behavior is modeled from the ant colony optimization algorithm [12]. This kind of agent accounts for an individual learning aspect similar to preferential attachment, and has the capability to pursue self-organized behavior by considering its individual (local) activities. However, we show that restrict models of agent communication based exclusively on pheromone exchange require an extension to represent direct communication in order to generate PL data distributions.

To support and validate our claims, we have defined a typical scenario of social simulations in which agents are embedded in a geographic environment. Agent communication is modeled indirectly via pheromone sharing and directly via message exchange. The goal is to investigate in which circumstances the spatial events that are produced by the agents are known to follow a PL distribution.

2 Scenario Description and Agent Modeling

For investigating how spatial distribution of events following PL can be generated from multi-agents with ant-based behavior, we have defined an artificial scenario in which agents live in an environment represented by a digital map of a geographic region. The environment possesses a set of objects with no mobility, here called targets. The task of the agents is to choose a target, move towards it and make a commitment. Each commitment represents a generation of an event (since the event has a particular coordinate associated, here we call it a spatial event). At the end of the simulation, the events produced by the agents are examined to determine whether their spatial distribution follows a PL. Agents have one or more points of departure that we call gateways. Such points of departure represent places where they start the simulation before choosing a target and making a commitment. It is also assumed that, at the end of a predefined period (typically one simulation day), each agent returns to the initial gateway. Target selection is probabilistic and based on the target distance, pheromone sharing, and the agent's interaction with other agents. The agent's behavior for choosing a target is modeled following the ACO algorithm. More formally, the probability p_{cn} of an agent c, choosing a specific target of the environment n, is:

$$p_{cn} = \frac{[\tau_{cn}]^{\delta} \times [\varphi_{cn}]^{\beta}}{\sum_{\forall p \in N} [\tau_{cp}]^{\delta} \times [\varphi_{cp}]^{\beta}} \qquad (1)$$

Here, τ_{cn} represents the pheromone of target n, whereas N is the set of all targets the agent c considers while deciding where to make the next commitment. The other parameter, φ_{cn}, denotes a static value that represents the inverse of the Euclidian Distance between the current location of agent c and that of target n; we assume that the agent has the knowledge necessary to localize the closest exemplar target in the environment. The parameters δ and β in Eq. (1) determine the relative influence of τ_{cn} and φ_{cn}. The role of the parameters δ and β is the following: If $\delta=0$, the closest targets are more likely to be selected. If $\beta=0$, only the pheromone is at work (the spatial distribution of targets is not taken into consideration). Once the target is chosen, the agent will commit. The number of commitments per target represents how attractive (pheromone) these targets are.

Regarding the pheromone factor, τ_{cn}, this can be calculated as

$$\tau_{cn} = \rho \times (\tau_{cn}) + \Delta \tau_{cn} \qquad (2)$$

Equation (2) indicates that the learned experience factor, τ_{cn}, takes into account, a negative feedback factor ρ that represents the level of forgetfulness of the past experience of an agent with regard to target n; that is, the extent to which the agent considers the experience taken from the last day $(\Delta \tau_{cn})$ in relation to the experience it already has with regard to target n. Hence, given that τ_{cn} represents the level of confidence of an agent with a given target, we emphasize that with every day (or every fixed interval of time), the agents forget a little about their previous experience and are more influenced by the recent one. Equation (2) also shows that an initial succession of failures will lose its influence over the agent's lifetime. It is worth noticing here that, in terms of ant-based swarm systems, the parameter ρ is used to represent a pheromone evaporation rate. Conversely, in our model, this parameter represents the rate at which agents forget past commitments.

2.1 Introducing Direct Inter-agent Communication

With the purpose of comparison to the ACO basic model described previously, we have augmented it to capture the direct interaction between agents in which the agent peer's opinion with respect to a given target is shared. In this model, here called social model, the learned experience τ_{cn} is given as a function of two terms, τ_{own} (the same as Eq. 2) and τ_{social} , representing, respectively, the private and collective experiences with respect to target n. The importance of this social learning factor is controlled by the parameter µ in Eq. (3). In this model, all contacts of c tell separately what they think about target n. Such hints are then aggregated. The value of the hints represents the consensual judgment the acquaintances of c have with respect to the target being considered (n). It is assumed that each agent has a set of acquaintances that make up part of its social network. At the end of a period, via direct message, the friends (those directly connected in the friendship graph) exchange information about their experiences in regard to targets already selected for commitment. Such experience is represented by the probability of choosing a particular target by the friends (see Eq. 4) and becomes a factor to be taken into account while the agent does the probabilistic target selection.

The network topology of the social network is also an important factor in this process, because it determines the speed at which information is propagated between friends. For instance, in a totally connected network, for each step of the simulation, a particular agent receives information from all other agents.

$$\tau_{cn} = \mu \times (\tau_{own}) + (1 - \mu) \times (\tau_{social})$$
(3)

Where:

$$\tau_{social} = \sum_{\forall k \in S(c)} (p_{kn})$$
(4)

Note that the social network related to the agent c is indicated as S(c) in Eq. (4).

3 Methodology of Evaluation

In what circumstances of the scenario and model of agents described above does one verify that the data produced by the agents (computed from each commitment) follow a Power Law? Is direct communication between agent members a predominant factor for this, or is only the information of distance and pheromone required? Does the initial distribution of agents in the environment cause any impact on the generation of data? Would the variation of the diver parameters of the ACO model generate events that are spatially distributed following a PL? To answer these questions and better understand the phenomenon here investigated, we performed the experiments described below. For a quantity of one hundred agents and simulations that were repeated three times, we varied each of the following parameters: distribution of the agents in gateways, the strategy of choice of targets by the agents, the evaporation factor, and the relative importance of the distance and the pheromone.

Aimed at analyzing the influence of the initial configuration of the agents with respect to the emergence of a PL, we proposed the distribution of agents in gateways in three different ways:

    i)   Random distribution;
    ii)   All agents in one single gateway;
    iii) PL distribution of agents.

As for the strategy of target selection, four manners were proposed:

i)   The agent takes into account only the distance to the target;
ii) The agent takes into account only the pheromone of the target;
iii) The agent takes into account the distance to the target and the pheromone of the target (Original ACO);
iv) The agent takes into account the distance, the pheromone and the direct communication of other agents that are part of the same social network. We give the same importance to the individual experience and to the collective information (coming from the others) ($\mu=0.5$).

The value of the forgetting factor (ρ) that we used as a standard was 0.54, and was varied to 0.14 and 0.94. For each possible combination of these previous items, we executed three simulations with 100 agents and 177 targets randomly distributed in the environment. For each one of these combinations, the data produced by the simulations (number of commitments) were plotted on doubly logarithmic axes in order to verify the fitting with a line (R^2) computed from linear regression. The estimation of the exponent value for the PL was also made from the linear regression by the identification of slope α of R^2. At the end of each set of five simulations, the averages of R^2 and the exponents (α) were computed. Note that for R^2, we considered the ideal value as between the interval [0.9, 1]. As for the value of α (exponent of the PL), we considered that a value greater than -1 is indicative of a slope that is representative of a PL, while values in the interval [-2, -4] are the most commonly found [7] and [8].

4   Results

Como citado na seção acima, para cada conjunto de parâmetros do modelo executamos simulações variando o número de tics da simulação a fim de analisar o impacto da variação do tempo da simulação. Veja a figura abaixo que contém os resultados a cerca da simulação usando ACO com parâmetros sugeridos na literatura.

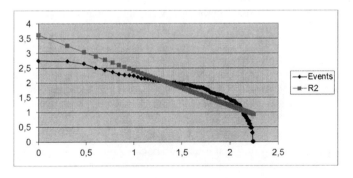

Fig. 1. Plot Log-Log using original ACO

Nesta figura percebemos a linha azul que representa o numero de eventos numa escala logarítmica. A linha rosa representa os valore ideal que os dados deveriam apresentar. O quão próximo a linha azul estiver da rosa representa o quão bom os dados seguem uma lei de potência. Esta proximidade é medida através do valor de R^2. Mas não é suficiente os valores estarem próximos da reta R^2, é necessário também que esta reta tenha um ângulo de inclinação específico, a inclinação da reta R^2 é medida através do parâmetro α. Neste caso a inclinação da reta R^2, ou seja, o valor de α, é de -1,7. E o valor de R^2 é de 0.71. A fim de congregarmos o maior número de informações possíveis em um único gráfico criamos o seguinte gráfico, que apresenta os valores de R^2 e α para cada combinação de parâmetros, onde cada ponto no gráfico representa uma quantidade especifica de tics da simulação e está associado a um segmento de reta de uma cor diferente representa uma distribuição de agentes nos gateways distinta.

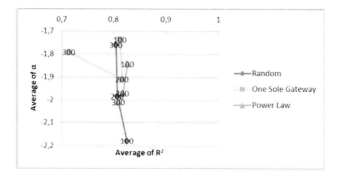

We will present the results of tests following the order of the target selection strategies shown in the previous section. For each target selection strategy, we run three simulations varying the number of tics (100, 200 and 300 tics) and ways of distributing agents in the gateways (random, one gateway and PL). The results can be seen in graphs where each coordinate (R^2, α) refers to the average value obtained for these parameters at the end of the three simulations in the number of tics determined. For example, in Figure 1, we see that for cases in which agents choose their targets based solely on the distance thereto, in the average of the three simulations conducted with agents randomly distributed in the gateways, the values of R^2 and α were (0.82, -0.13) for simulations with 100 tics, (0.80, -0.1) for simulations with 200 tics, and (0.77, -0.07) for simulations with 300 tics.

Inicialmente the agents have a partial view of the environment, i.e., when they do not examine all of the targets while trying to carry out an event because they can only see those that are within their view, we do not observe the expected result. In the other simulations, we used the degenerate case in which agents have full view, i.e., they can evaluate all of the targets in the choice of a target.

Na primeira fase das simulações, the agents consider only their distance to the target in the selection thereof, even though they carry out events in places that show a distribution close to a straight line, they have very low values of α, entre -0.1 e -0.2. This is a straight line without any slope that cannot be characterized as a PL.

By running the simulations with agents choosing the target only by the pheromone, one can see in Figure 1 that the value of α improves considerably, but the R^2 values are still not significant. It is noteworthy that with the increase in the number of tics, the value gives indications of distancing itself from the desired range.

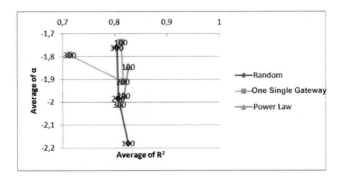

Fig. 2. Results using only the pheromone

Também executamos extreme situations of the ACO model. Primeiramente, the value of δ was zero and in the other one the value of β was zero. When the two values are taken into account equally (δ = 1 and β = 1), the values of R^2 and α show slight improvements, although not significant in terms of fit to a PL.

Different proportions for the values of δ and β were tried, but the results gave no indication that there would be any significant difference. We noted that when we give more importance to the pheromone than the distance, the value of R^2 is very poor, no matter what type of exit of agents from the gateways; when we prioritize distance, the results are not much different from when we give equal importance to both parameters.

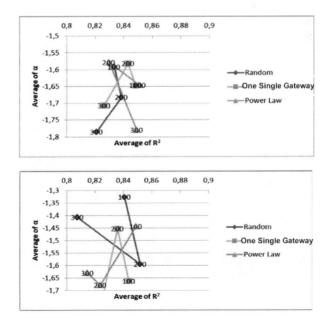

Fig. 3. Results varying the time to return to the Gateways, with values respectively equal to 5 and 15

Another parameter that deserves analysis was the rate of evaporation ρ of the pheromone used in previous simulations, which was $\rho = 0.54$. The variation of this parameter to a higher value (0.94) and a lower value (0.14) also showed no significant improvements regarding the formation of PL.

We also varied the parameter "return time to gateway," representing the range of tics in which the agent returns to the gateway from which it left. The default value used is 1 (one), because we understand that the agent only commits one event per day, so each tick represents one day in the life of the agent. We can analyze that the variation of this parameter did not produce a considerable improvement in the results, and that the results of all three types of distribution in gateways are very similar to one another, both for the value 5 and for the value 15 of the parameter.

In general the results of the simulations gave an indication that the ACO model, although typical characteristics of preferential attachment are embedded in its model, is not sufficient to generate events that are distributed in space following a PL.

Since the ACO model with indirect communication was not able not reach values close to what we stipulated as a characteristic of PL distribution, we then evaluated the ACO model with direct communication. In this case, it was assumed that all of the agents that are part of the simulation and are able to carry out a commitment were part of a single social network represented by a "small world" graph.

Figure 4 presents the results of the modified ACO, using the social communication of the agents, i.e., at the time of selecting targets, the agents take into account the pheromone of that target, their distance to such target, and the opinion of its direct "friends" (direct links on the agent relationship graph) about that target. We perceive that social communication significantly improves the results obtained thus far, in both parameters for assessing the outcome, R^2 and α. Some results present an extremely high degree of relationship with the PL. In Figure 4 one can see how the distribution of the events generated in the simulation behaved spatially.

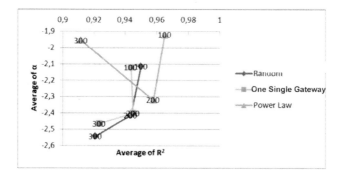

Fig. 4. Results using ACO algorithm with social communication

5   Discussion

In general, the distribution of the agents in gateways was not shown to be a decisive factor for the events generated to follow a PL. The configuration in which the agents leave a single gateway shows high variability, because it depends heavily on the location of the targets.

The most relevant conclusion comes from the fact that increasing the perception of the agents by introducing direct communication among them was an important factor for improving the outcome, regardless of the configuration adopted. This reinforces the hypothesis that the partial view of the agents and the mechanism of indirect communication thereof is not enough to perceive the benefits of all of the targets and thus the distribution of a Power Law is not shown so clearly.

6 Related Work

Power Law in general has been a subject of great interest in the community, both in generation as well as the methods for analyzing experiments conducted.

Multi-agent modeling of phenomena that generate PL has been investigated from the viewpoint of criticality. In that context, the modeling of events that accumulate in a certain place and at a certain critical point are disturbed, and by diffusion or propagation were studied in [10]. In this context, the modeling of the spatiality of the events depends on the locations of the targets. The analysis in this article requires no information about the location of targets. The focus here was on the behavior of the agents.

Under this focus, it is most common to find generators that – in a centralized manner – have a pre-programmed function generator. Barabasi [4], for example, presents a simple method to generate networks where the connections form a PL. A probabilistic function favors the choice of nodes that already have more connections, which characterizes what he himself coined "preferential attachment."

In another context, that of computer science, a more ingenious solution was proposed by [5]. The authors propose a generator that can generate not only a PL distribution, but other distributions as well. This generator, called R-MAT (Recursive Matrix) generates graphs with a community structure that have a small diameter through recursive calls to a method that assigns different probabilities of insertion of graph edges to the adjacency matrix quadrants of the graph nodes.

Both studies generate good results regarding Power Law match. However, unlike this work, they use centralized strategies, which do not occur commonly in nature.

In another different context, terrain coverage algorithms have been used in several applications as web crawling and automated harvesting. Even though with different goal of ours, they are worth mention because in several of these situations the number of times a places are visited follows a PL as shown in [13].

7 Conclusion

In this paper we have investigated the generation of spatial events that follow a PL shedding light on how this can be obtained via multi-agent models. Particularly, we investigated models in which the agent behavior is based on a preferential attachment strategy existing in the Ant Colony Optimization heuristic. The simulation scenario we created allowed the interpretation of main circumstances that can affect the generation of PL without knowledge about the target distribution.

The ACO heuristics alone is not enough to generate PL distribution in the terms as they are encountered in the literature. The introduction with direct communication mimicking the interaction that exists between agents of a same community has shown to be a crucial factor. In this case, spatial events present a strong fit with PL, particularly with Zipfian distributions.

References

1. Lewis, T.G.: Network Science. Wiley, Chichester (2009)
2. Clauset, A., Shalizi, C.R., Newman, M.E.: Power-law distributions in empirical data. arXiv:0706.1062v1 (2007)
3. Girvan, M., Newman, M.E.: Community structure in social and biological networks. Proceedings of the National Academy of Sciences 99(12), 7821–7826 (2002)
4. Barabasi, A., Albert, R.: The emergence of scaling in random networks. Science 15:286(5439), 509–512 (1999)
5. Chakrabarti, D., Faloutsos, C.: Graph Patterns and the R-Mat generator. In: Cook, D., Holder, F. (eds.) Mining Graph Data, pp. 65–95. Wiley, Chichester (2007)
6. Engelmore, R., Morgan, A.: Blackboard Systems. Addison-Wesley, Reading (1986)
7. Dorogovtsev, S.N., Mendes, J.F.F.: Evolution of Networks. Oxford University Press, New York (2003)
8. Albert, R., Barabasi, A.L.: Review of Modern Physics 74, 47 (2002)
9. Miller, J., Page, S.: Complex Adaptive Systems: An Introduction to Computational Models of Social Life. Princeton University Press, Princeton (2007)
10. Bak, P.: How Nature Works. Springer, New York (1996)
11. Sichman, J., Antunes, L.: Multiagent-based Simulation VI. Springer, Heidelberg (2005)
12. Dorigo, M., Stulze, T.: Ant Colony Optimization. MIT Press, Cambridge (2004)
13. Mason, J., Menezes, R.: Autonomous algorithms for terrain coverage: Metrics, classification and evaluation. In: Proceedings of the IEEE World Congress on Computational Intelligence, pp. 1641–1648. IEEE Computer Society, Hong Kong (2008)

Using Digraphs and a Second-Order Markovian Model for Rhythm Classification

Debora C. Correa[1], Luciano da Fontoura Costa[1,3], and Jose H. Saito[2]

[1] Instituto de Física de São Carlos, Universidade de São Paulo, São Carlos, SP, Brazil
deboracorrea@ifsc.usp.br, luciano@ifsc.usp.br
[2] Departamento de Computação, Universidade Federal de São Carlos, São Carlos, SP, Brazil
saito@dc.ufscar.br
[3] Instituto Nacional de Ciência e Tecnologia para Sistemas Complexos, Centro Brasileiro de Pesquisa Física, Rio de Janeiro, RJ, Brazil

Abstract. The constant increase of online music data has required reliable and faster tools for retrieval and classification of music content. In this scenario, music genres provide interesting descriptors, since they have been used for years to organize music collections and can summarize common patterns in music pieces. In this paper we extend a previous work by considering digraphs and a second-order Markov chain to model rhythmic patterns. Second-order transition probability matrices are obtained, reflecting the temporal sequence of rhythmic notation events. Additional features are also incorporated, complementing the creation of an effective framework for automatic classification of music genres. Feature extraction is performed by principal component analysis and linear discriminant analysis techniques, whereas the Bayesian classifier is used for supervised classification. We compare the obtained results with those obtained by using a previous approach, where a first-order Markov chain had been used.Quantitative results obtained by the kappa coefficient corroborate the viability and superior performance of the proposed methodology. We also present a complex network of the studied music genres.

1 Introduction

The continuous growth in size and number of online music databases has required accurate and faster tools for researches for music content analysis, retrieval and description. Applications involving interactive access and music content-based queries have received much attention due to the large amounts of online music data. Therefore, recent research approaches have focused on the development of techniques to analyse, summarize, index, and classify music data. In this context, music genres provide particularly interesting descriptors, once they have been widely used for years to index and organize music collections. In addition, music genres are important since they reflect the interplay of cultures, trajectories and interactions of artists and market strategies. They are also of great interest because they summarize some common characteristics (patterns) in music artworks. Despite their vast use, music genres have fuzzy and controversial boundaries and taxonomies [12,13]. Even extensively used terms like *rock*, *jazz* and *blues* are not firmly defined. As a consequence, the development of a clear

L. da F. Costa et al. (Eds.): CompleNet 2010, CCIS 116, pp. 85–95, 2011.
© Springer-Verlag Berlin Heidelberg 2011

taxonomy is a challenging problem, which corroborates the fact that the automatic classification of music genres is not a trivial task.

As commonly happens in many problems involving pattern recognition, the task of automatically classifying music genres can be subdivided into three main steps: representation, features extraction, and classification. Music information can be characterized through acoustic signals or symbolic representation, such as MIDI, where each note is an event described in terms of pitch, duration, strength and start and end times [3]. Music features can be used to quantify aspects related to melody, timbre, rhythm and harmony. In the literature, there are many previous works dealing on automatic classification of music genres [9,11].

An extention of an previous approach to automatic genre classification is proposed in the current work. Similarly to that previous approach, our principal objective here is to classify the music genres features related to their rhythmic patterns. Rhythm is a collection of patterns produced by notes, which are different in duration, stress and pause, arranged with temporal regularity. However, there is no agreement about the definition of rhythm [13]. Besides been simpler and easier to manipulate than the other aspects of music content in the artworks, the rhythm can intrinsically and intuitively characterize the musical genres. Through specific rhythmic patterns produced by their notes we can immediately distinguish, for example, waltz and salsa music. Besides, the rhythm is not dependent on the arrangements. Many previous works address automatic classification of music genres by using rhythms features [1,8].

In a previous work [6], the problem of automatic genre classification was explored in terms of rhythmic features. Actually, the authors proposed a first-order Markov chain to model rhythmic patterns. In the current study we complement and extend that analysis by considering a second-order Markov chain together with additional features. We show that the consideration of second-order time dependencies can lead to a more accurate model of the rhythmic patterns, substantially improving the classification results. The samples (artworks) were obtained in MIDI format, from which several rhythmic related measurements were computed. Then, we distinguish the music genres through their rhythmic patterns, taking into account the occurrence of sequence of events. The rhythmic patterns are represented as digraphs. Second-order transition probability matrices were computed from the rhythms of each piece, producing a second-order Markovian model.

We use the Bayesian classifier for the supervised classification approach. The classifier receives as input the extracted features and produces as output the most likely genre. We used the same dataset as in [6], which contains samples from four well-known genres: blues, *mpb* (Brazilian popular music), reggae and rock. We show a complex network built from the features of these four music genres.

The remaining of this paper is organzed as follows: section 2 presents the details concerning the methodology; section 3 describes the experiments; and section 4 presents the conclusions.

2   Methodology

Here we describe the dataset and proposed methodogy in the current study.

2.1   Data Description

As one of the purposes of this work is to extend a previous study and compare the results, we used the same database considering four music genres: blues, *mpb*, reggae and rock. These genres are broadly known and represent distinct tendencies. Furthermore, respective music samples (artworks) can be easily found in Web collections. The same seventy samples of each genre were selected and represented in MIDI format. MIDI is an event-like format that can be viewed as a digital music score in which the instruments are separated into voices [10].

With the purpose of editing and analysing the MIDI scores, we applied the Sibelius software for music notation (http://www.sibelius.com). It is known that the percussion is inherently suitable to express the rhythms of the artworks. Therefore, the voice related to the percussion was extracted from each sample. Then, it is possible to analyse the elements involved in the rhythm. The free Midi Toolbox for Matlab computing environment was used to perform such analysis [7]. Actually, a MIDI file is transformed into a matrix representation of note events, whose columns refer to information such as onset (in beats), duration (in beats), MIDI channel, MIDI pitch, among others; while its rows represent the individual note events, that is, the notes are described in terms of pitch, duration, and so on.

Following the same idea as in [6], we consider the occurrence of sequence of events by rhythmic notations, which are referred as the note duration (in beats). In [6], a digraph was built considering the duration of the notes through the sequence in which they occurred in the artwork. In this case, each vertex represent a rhythmic notation (such as half note, quarter note, eighth note, sixteenth note, and so on) while the edges indicate the subsequent note. Therefore, a first-order Markov chain was applied. Nevertheless, we suggest that a second-order Markov chain can better capture the dynamics and dependencies of the rhythms, because it provides an extended context at each time. Thus, we propose to consider not only the dependency of the current note on the previous one, but instead, the dependency of the current note on both the previous notes. In this context, the vertices can represent a rhythmic notation or a pair of rhythmic notations. For example, the existence of an edge from vertex i, represented by an sixteenth and an eighth note, to a vertex j, represented by a quarter note, indicates that the sequence sixteenth note - eighth note - quarter note occurred at least once. The thicker the edge, the larger is the strength between the two nodes involved. In orther words, this sequence occurs more frequently in the piece.

Table 1 is an illustration of a second-order transition probability matrix, obtained through the durations sequence of the first bars of the music *From Me To You* by The Beatles, presented in Figure 1(a). The digraph representation of this piece is respectively illustrated in Figure 1(b). We also show the digraph (Figure 1(c)) built to express the first-order transition probability matrix on the durations sequence of the same melody. It is clear that the second-order digraph is substantially richer, providing more information regarding the rhythm dynamics.

A second-order digraph is built for each artwork in the dataset taking into account the duration of the notes, i.e., analysing the sequence in wich they occurr in each artwork. As the instrumentation is not been investigated in this work, and the percussion consists of different simultaneous instruments, when two or more events happen in the

Table 1. Example of a second-order transition probability matrix

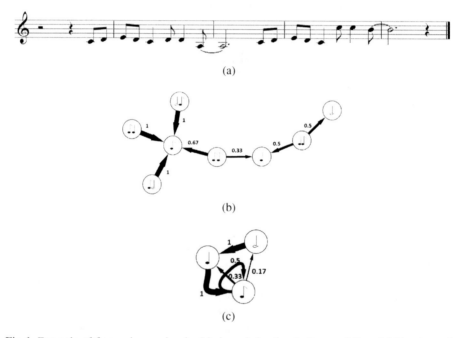

(a)

(b)

(c)

Fig. 1. Example of first and second-order Markov chains for rhythm modeling. (a) First bars of the melody *From Me To You* by The Beatles. (b) Digraph representation of the second-order transition probability matrix in Table 1. (c) Digraph representation of the first-order transition probability matrix for the melody in (a).

same beat, the median duration of them is obtained in the same was as in [6]. Digraphs examples are shown in Figure 2. Figure 2(a) shows a blues sample represented by the music *Looking Out The Window* by Stevie Ray Vaughan. An MPB sample, represented by the music *Desalento* by Chico Buarque, is depicted in 2(b). Figure 2(c) shows a reggae sample, namely the music *Three Little Birds* by Bob Marley. Lastly, Figure 2(d) illustrates a rock sample, corresponding to the music *She Loves You* by The Beatles.

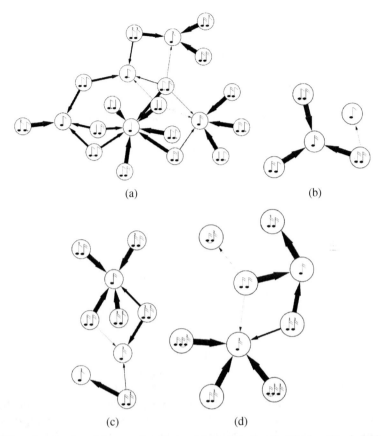

(a)   (b)

(c)   (d)

Fig. 2. Digraph examples of four music samples. (a) *Looking Out the Window* (Stevie R Vaughan); (b) *Desalento* (Chico Buarque); (c) *Three Little Birds* (Bob Marley); (d) *She Loves You* (The Beatles).

2.2 Feature Extraction and Analysis

In most pattern recognition systems, feature extraction is a fundamental step since it directly influences the performance of the system. The patterns (or samples) are represented by vectors within a d-dimensional space, where d is the total number of features or attributes of each sample. Generally, the principal aim in discrimination problems is to maximize class separability, that is, pattern vectors belonging to different classes should occupy compact and distinct regions in the feature space. Consequently, reliable good features must be selected to allow such discrimination.

The previous section described how a digraph can be created for each of the 70 samples in each genre considering the second-order Markov chain. We used the same 18 possible rhythmic notations as in [6]. The number 18 came from the exclusion of some rhythmic notations that hardly ever happen. This aspect is important because it increases the classification accuracy, since it reduces the data dimensionality, avoiding features that do not significantly contribute to the analysis.

In order to provide an appropriate comparison of our results and the results in [6], we divided the features into the two following groups:

Feature Matrix A corresponds to the features related to the weight matrix W. The columns of matrix W represent the 18 possible notations, while the rows represent all possible combinations of these rhythmic notations in pairs. More specifically, these pairs express all the sequence of the two past rhythmic notations occurred before the current rhythmic notation in analysis. As commented previously, each element in W, w_{ij}, associates a weight in the connection from vertex i (indicating the two past rhythmic notations), to vertex j (indicating the current rhythmic notation). The weight expresses the frequency that the past rhythmic notations were followed by the actual rhythmic notation in the sample, just as exemplified in Table 1. After analysing all the samples in the dataset, we verified that 107 pairs of durations would be necessary to represent all possible sequences. As a result, the weight matrix W for each sample has 167 rows and 18 columns. In order to obtain the final feature matrix, each weight matrix W was reshaped by a 1 x 3006 feature vector, resulting in a final feature matrix A of 280 rows and 3006 columns (the attributes).

Feature Matrix B comprises the measurements extracted directly from the digraphs. Vertices in the columns of the weight matrix W do not have outgoing edges, while vertices in the rows do not have incoming edges. For each one of the samples in the dataset, the following measures were computed: the median out-degree, calculated over all vertices corresponding to rows of matrix W; the median in-degree, calculated over all vertices corresponding to columns of matrix W; the total median degree (median out-degree + median in-degree); standard deviation of the out-degree; standart deviation of the in-degree; maximum value of the out-degree; maximum value of the in-degree; the median out-strength; the median in-strength; standard deviation of the out-strength; standard deviation of the in-strength; maximum value of the out-strength; maximum value of the in-strength. Therefore, the final feature matrix B has 280 rows and 14 columns.

Classification performance can be usually improved by the normalization of the features (zero mean and unit standard deviation) [14]. Once the normalized features are available, the structure of the extracted rhythms were analysed with two different approaches for features analysis: PCA and LDA. Principal Components Analysis (PCA) and Linear Discriminant Analysis (LDA) are two techniques widely used for feature analysis [14]. Both approaches apply geometric transformations (rotations) to the original feature space, generating new features that are linear combinations of the original ones. PCA seeks a projection that creates new orthogonal uncorrelated features. In other words, it removes the features redundancy, allowing dimensionality reduction. LDA, on the other hand, aims for a projection that best separates the data, improving the classification performance.

2.3   Classification Methodology

We used the Bayesian classifier through discriminant functions as the supervised classification approach. Based on the Bayesian Decision Theory, the Bayesian classifier uses conditional densities and prior probabilities to assign each object to the most likely class[14]. It defines an optimum classification rule, since it ensures minimum probability of misclassification, provided that the respective class densities are properly estimated.

A quantitative criterion is usually taken to objectively evaluate the performance of the supervised classification. Most approaches only use the classification error and the obtained accuracy. We also adopted the Cohen Kappa Coefficient [4] as the quantitative criterion, since it has good statistical properties (e.g., be asymptotically normal) and can be directly calculated from the confusion matrix [5].

In the applied classification tasks, re-substitution means that all objects from each class were used as the training set (to estimate the parameters) and all objects were used as the testing set. Hold-out 70%-30% means that 70% of the objects from each class were used as the training set and 30% (different ones) for testing. Finally, in hold-out 50%-50%, the objects were separated into two groups: 50% for training and 50% for testing.

Figure 3 presents the summarized flow diagram of the proposed methodology.

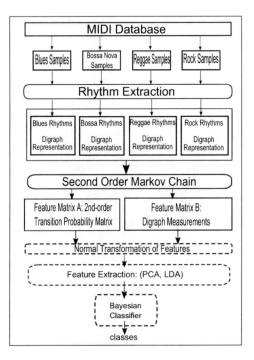

Fig. 3. Flow chart of the proposed methodolody

3   Experiments and Discussion

From the PCA and LDA components, we observed that the classes are completely overlapped. In the case of PCA, at least 30 components were necessary to preserve 75% of the data variance, a value often acceptable in pattern recognition tasks. Therefore, we

Table 2. PCA kappa and accuracies for feature Matrix A

	Kappa 1st-order	Accuracy	**Kappa 2nd-order**	**Accuracy**	**Performance**
Re-Substitution (49)	0.61	70.72%	0.82	87%	Almost Perfect
Hold-Out (70%-30%) (20)	0.33	50%	0.32	49%	Fair
Hold-Out (50%-50%) (10)	0.32	48.57%	0.35	51%	Fair

Table 3. PCA kappa and accuracy for feature Matriz B, and feature Matrix A + feature Matrix B - Re-Substitution

	Kappa	**Accuracy**	**Performance**
Matrix B (15)	0.47	60%	Moderate
Matrix A + Matrix B (56)	0.9	93%	Almost Pefect

Table 4. Confusion Matrix for the classification with the combination of feature Matrix A and B (kappa=0.9)

	Blues	*MPB*	Reggae	Rock
Blues	67	0	0	0
MPB	0	63	0	3
Reggae	3	7	70	7
Rock	0	0	0	60

can reduce the data dimensionality from 3006-D to 30-D. However, the ideal number of components to achieve suitable classification results can vary, according to the classification task. The numbers within brackets in front of each classification task in Tables 2 and 3 indicate the ideal quantity of components to achieve the highest value of kappa 2nd-order.

Tables 2 and 3 present the classification results obtained by the Bayesian classifier (over the PCA features), in terms of the kappa coefficient, the respective performance according to its value [5], and the accuracy of the classification. We compare these results with those obtained by the previous approach in [6], that uses a digraph built from a first-order Markov chain. Using a second-order Markov chain can significantly improve the classification results. The classification performance was not satisfactory for the Hold-Out situations, although there was a slight improvement in the case of Hold-Out 50%-50%. As PCA is a unsupervised approach, the performance of estimation of the covariance matrices is strongly degraded due to the small sample size problem. This can be minimized with a larger dataset. The combination of feature matrix A and B provided 93% of correct classifications. Keeping in mind that the problem of automatic classification of music genres is a nontrivial task, and that our focus is only on the rhythm analysis (drastically reducing computational costs), the obtained PCA results are quite interesting and substantial.

Table 4 presents the confusion matrix over classification with the combination of feature Matrix A and B. Table 5 shows the classification results obtained by the Bayesian classifier over the LDA features. The proposed approach again increased the classification results. LDA is an interesting technique because it uses all discriminative informa-

Table 5. LDA kappa and accuracies for feature Matrix A

	Kappa 1st-order	Accuracy	**Kappa 2nd-order**	**Accuracy**	**Performance**
Re-Substitution	0.62	71.78%	0.8	85%	Substantial
Hold-Out (70%-30%)	0.57	67.5%	0.87	90%	Almost Perfect
Hold-Out (50%-50%)	0.6	70%	0.84	87.85%	Almost Perfect

Table 6. LDA kappa and accuracy for feature Matrix A + feature Matrix B

	Kappa	**Accuracy**	**Performance**
Re-Substitution	0.85	89%	Almost Perfect
Hold-Out (70%-30%)	0.92	93.75%	Almost Perfect
Hold-Out (50%-50%)	0.87	90%	Almost Perfect

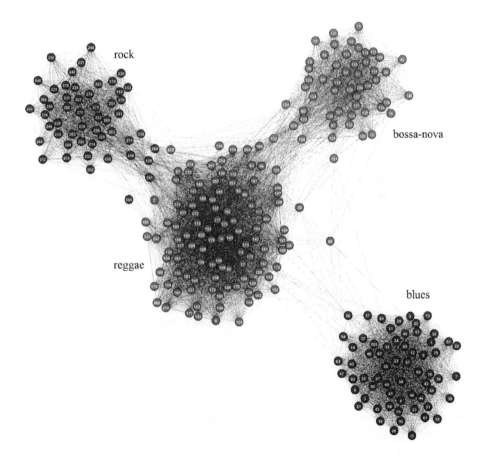

Fig. 4. The complex network of the four music genres. Linked vertices are attracted to one another and non-linked vertices are pushed apart. This process led to four communities of music genres.

tion available in the feature matrix, but it is restricted to computing only $C - 1$ nonzero eigenvalues, where C is the number of class. As a consequence, only three components (three new features) can be obtained. Different from PCA, LDA provided suitable results in the Hold-Out situations, reflecting the supervised characteristic of it.

Figure 4 presents an example of a complex network of music genres with respect to the rhythmic features. The vertices represent the music artworks in the dataset. Each vertex has a feature vector associated to it, which, in this case, corresponds to the LDA features of the respective artwork. The weight of the connection of two vertices is $1/d$, where d is the Euclidean distance between them. At first, therefore, all vertices are connected to each other. However, in order to provide an suitable visualization of the groups, a threshold was used to eliminate some edges with small weights, reflecting the more distant vertices. The network was visualized by using the free software Gephi (`http://gephi.org/`). The layout was obtained through the algorithm "Force Atlas" in Gephi. This is a force-based algorithm whose principle is to attract linked vertices and push apart non-linked vertices. For community detection, Gephi implements the Louvain method [2]. We observed that the reggae class is the one that most overlap with the others, mainly with the class rock, while blues is the most separated class. In addition, some artworks are placed in classes that are not their original ones, indicating that their rhythmic patterns are uncommon.

4   Conclusions

In order to extend and complement a previous study [6], a second-order Markov chain approach has been explored here in the context of automatic music genre classification through rhythmic features. Four music genres are used: blues, *mpb*, reggae and rock. The rhythmic patterns were obtained from MIDI files and modeled as digraphs, depending on the respective second-order transition probability matrix, which comprises the first group of features. The second group involves measures obtained directly from the digraphs. We used PCA and LDA approaches for feature extraction, while the classification was applied by the Bayesian classifier.

We verified that Matrix B, with only 14 attributes could lead to considerable classification performance in a much lower dimensional space. The combination of the two feature matrices lead to 93% of correct classifications, with kappa = 0.9. In respect to the statistical significance of kappa, this result and the one obtained only with feature matrix A (kappa=0.82), are different.

We observe that the rhythms are substantially complex, yielding four overlapping classes, since there are many samples with similar feature vectors common to more than one of those groups. Even considering an extended context at each time to capture the dynamics of rhythmic patterns, the dimensionality of the problem remains high, implying that many features may be required to separate them. However, the use of a second-order Markov model led to substantially better results. In fact, higher-order Marvov chains tend to incorporate an extented rhyhmic structure, rather than only analyse straightforward sequences as in the case of first-order models.

Additionally, even music experts can have difficulties to distinguish between the genres, since generally a song can be assigned to more than one genre. If only one aspect of the rhythm is been considered, this difficulty increases significantly. The intensity of the beat, which is a fundamental aspect of the rhythm is not been analysed, i.e., analysing rhythm only through rhythmic notations, as proposed here, could be hard even for humans experts. All these issues corroborate the viability of the presented methodology, once suitable results were obtained within a limited information content, which reduces the computational costs and simplifies the understanding of the features analysis.

It is also clear that the proposed methodology requires some complementation. It would be interesting to extract more measurements from the rhythms, specially the intensity of the beats. The use of different classifiers as well as their combination could also be investigated. Nowadays, most MIDI databases are still single labeled (each sample belongs to one single genre). However, the authors are investigating multi-label classification approaches, since it better reflects the intrinsic nature of the dataset.

Acknowledgement. Debora C Correa thanks FAPESP (2009/50142-0) for financial support and Luciano da F. Costa is grateful to CNPq (301303/06-1 and 573583/2008-0) and FAPESP (05/00587-5) for financial support.

References

1. Akhtaruzzaman, M.: Representation of musical rhythm and its classification system based on mathematical and geometrical analysis. In: Proceedings of the International Conference on Computer and Communication Engineering, pp. 466–471 (2008)
2. Blondel, V.D., Guillaume, J., Lambiotte, R., Lefebvre, E.: Fast unfolding of communities in large networks. Journal of Statistical Mechanics: Theory and Experiment (10), 10,008 (2008)
3. Cataltepe, Z., Yasian, Y., Sonmez, A.: Music genre classification using midi and audio features. EURASIP Journal on Advances of Signal Processing, 1–8 (2007)
4. Cohen, J.: A coefficient of agreement for nominal scales. Educational and Psychological Measurement 20(1), 37–46 (1960)
5. Congalton, R.G.: A review of assessing the accuracy of classifications of remotely sensed data. Remote Sensing of Enviroment 37, 35–46 (1991)
6. Correa, D.C., Costa, L.d.F., Saito, J.H.: Musical genres: Beating to the rhythms of different drums. New Journal of Physics 12(053030), 1–38 (2010)
7. Eerola, T., Toiviainen, P.: MIDI Toolbox: MATLAB Tools for Music Research. University of Jyväskylä, Jyväskylä (2004), http://www.jyu.fi/musica/miditoolbox/
8. Gouyon, F., Dixon, S.: A review of automatic rhythm description system. Computer Music Journal 29, 34–54 (2005)
9. Li, T., Ogihara, M., Shao, B., Wang, D.: Machine Learning Approaches for Music Information Retrieval. In: Theory and Novel Applications of Machine Learning, pp. 259–278. I-Tech, Vienna (2009)
10. Miranda, E.R.: Composing Music With Computers. Focal Press (2001)
11. Mostafa, M.M., Billor, N.: Recognition of western style musical genres using machine learning techniques. Expert Systems with Applications 36, 11,378–11,389 (2009)
12. Pachet, F., Cazaly, D.: A taxonomy of musical genres. In: Proc. Content-Based Multimedia Information Acess (RIAO) (2000)
13. Scaringella, N., Zoia, G., Mlynek, D.: Automatic genre classification of music content - a survey. IEEE Signal Processing Magazine, 133–141 (2006)
14. Webb, A.R.: Statistical Pattern Recognition. John Wiley & Sons Ltd., Chichester (2002)

The Effect of Host Morphology on Network Characteristics and Thermodynamical Properties of Ising Model Defined on the Network of Human Pyramidal Neurons

Renato Aparecido Pimentel da Silva[1], Matheus Palhares Viana[1],
and Luciano da Fontoura Costa[1,2]

[1] Instituto de Física de São Carlos, Universidade de São Paulo,
PO Box 369, 13560-970, São Carlos, SP, Brazil
rapsilva@ursa.ifsc.usp.br, vianamp@gmail.com
[2] Instituto Nacional de Ciência e Tecnologia para Sistemas Complexos,
Centro Brasileiro de Pesquisa Física,
R. Dr. Xavier Sigaud 150, 22290-180, Rio de Janeiro, RJ, Brazil
luciano@ifsc.usp.br

Abstract. The question about the effect of the host (node) morphology on complex network characteristics and properties of dynamical processes defined on networks is addressed. The complex networks are formed by hosts represented by realistic neural cells of complex morphology. The neural cells of different types are randomly placed on a 3-dimensional cubic domain. The connections between nodes established according to overlaps between different nearest-neighbor hosts significantly depend on the host morphology and thus are also random. The influence of host morphology on the following network characteristics has been studied: edge density, clustering coefficient, giant component size, global efficiency, degree entropy, and assortative mixing. The zero-field Ising model has been used as a prototype model to study the effect of the host morphology on dynamical processes defined on the networks of hosts which can be in two states. The mean magnetization, internal energy and spin-cluster size as function of temperature have been numerically studied for several networks composed of hosts of different morphology.

1 Introduction

Networked systems appear everywhere. Examples include the social web, collaboration networks, power grids, railways and airline routes, to cite some. While in the first two mentioned cases the position of individual nodes is arbitrary or not defined at all, the same is not valid for the latter, *geographical* or *spatial networks* [4]. In such systems, the positions of the constituents restrain or even define their inter-connectivity: the distance separating two given nodes, for example, may determine the existence of a connection bonding them – in fact, for the majority of systems, this is the determinant aspect of inter-connectivity. However, other factors may be present, including global or spatial aspects such as viability (cost) and environment (e.g. fluvial transport); and individual aspects, including load (e.g. the Internet) and demand (transportation networks).

L. da F. Costa et al. (Eds.): CompleNet 2010, CCIS 116, pp. 96–107, 2011.

Neuronal networks provide probably the better example of interconnected system where another individual factor has fundamental importance: the *shape* of the individual hosts, namely the neurons. The complexity of dendritic arborization of each neuronal cell directly interferes on local connectivity, e.g. the number of immediate connections the neuron establishes. The generation of neuronal networks formed by realistic structures is essential for explore such aspects, and in the last years several computational frameworks aimed at this task have been developed, e.g. [9,10,22,13]. Such frameworks allow the construction of 3-dimensional neuronal networks of either realistic growing computer-generated cells or digital reconstructions of real neurons. Nonetheless, global network aspects [6,7] and dynamic processes other than cell growth taking place on such networks have been little explored, if not explored at all.

In the present article we study the individual host morphology impact on both overall network characteristics and dynamics, through the emulation of neuronal networks of 3-dimensional realistic digital reconstructions of human pyramidal cells. To address such effects, neuronal networks are generated by placing in contact individual neurons of two distinct shapes, chosen according to a criterion which takes into account several neuronal attributes including the number of bifurcations and branches, cell elongation, etc. The density of one of the shapes is allowed to vary, and global network attributes such as edge density, degree entropy and global efficiency are then evaluated for different values of such a density. In the case of dynamics, the ubiquitous Ising model is adopted. Widely explored in Statistical Mechanics, this model consists of a set of interacting elements in thermodynamic equilibrium, each one associated with either an up- or a down-aligned magnetic moment. The flipping of a magnetic moment changes the energy of the system, the order of the variation proportional to the immediate connectivity of the respective host, so that the overall topology and the dynamic behavior of the system are directly correlated. While more realistic aspects of the cells, such as the differentiation of neuronal processes – soma, axons, dendrites – are discarded in the connectivity model, the proposed approach provides interesting insights about the individual shape–overall dynamics interplay.

In the next section we address the digital reconstructions of human pyramidal cells. Afterwards, we briefly describe some important network measurements, which we believe are of interest for the analysis of neuronal networks. A brief explanation of the connectivity model, as well as the Ising dynamics employed follows. Finally, the results on both overall connectivity and dynamics are presented and discussed.

2   Neuronal Digital Reconstructions

Neurons are among the most intriguing natural structures. The anatomy and physiology of neuronal cells as currently known were firstly explored by the pioneer works of Santiago Ramón y Cajal and his contemporary Camillo Golgi, at the end of XIX Century. Both adopted the impregnation of a silver nitrate solution over *in vitro* nervous tissue to observe the neuronal cells. Cajal actually improved the method of Golgi through a technique he denominated "double impregnation". In this technique, a random fraction of neurons present in the tissue becomes impregnated, contrasting such cells against an yellowish background, a differentiation which turns the reproduction of neuronal cell shape possible. Cajal then reproduced the neurons via the *camera lucida* method. Camera lucida consists in using of a system of mirrors mounted along the microscope in order to project the framed image over a surface, which allows one to hand-trace the observed structures, thus generating a *2-dimensional* representation of the subject. The technique is still adopted by many experimental researchers.

Though the 2-dimensional tracing may be scanned into a computer to provide a digital reconstruction of the cell, such reconstructions of the neurons now can be directly obtained in digital format via a microscope–computer interface setup [3], thanks to advances on both microscopy and informatics. One of the benefits of the novel approach is the full *3-dimensional* reconstruction of the neuronal structure. The reconstruction obtained is generally segmented into cylindrical inter-connected compartments, and individual compartment information such as coordinates of end point, radius, compartment type (soma, axon, etc.) are then stored on a matrix. Such information has two important advantages: first, the matrix obtained can be stored on a file, allowing one to fully reconstruct the cell from the stored data; the second one is the fact that several neuronal *morphometric measurements* such as number of dendritic trees attached to soma, soma surface, number of dendritic branches or bifurcations present on the cell and so on, can be estimated from the matrix.

Today neuroscientists as well as researchers from related fields may obtain digital reconstructions of neuronal cells by downloading such data from online repositories, such as `http://neuromorpho.org` [2]. NeuroMorpho provides more than 5,600 digital reconstructions of neuronal cells, obtained from diverse Neuroscience laboratories. For each specific reconstruction, 20 morphometric measurements are given, as well as information such as the dataset the reconstruction belongs to, the species involved, the localization of the tissue in central nervous system, the slicing direction, etc. It is possible, for example, to obtain digital reconstructions of a specific neuron type, or belonging to a specific area of cerebral cortex.

For the present study we collect samples of human pyramidal cells, which represent about 37% of all the data available on NeuroMorpho repository. The neuron morphometries provide the information we need for numerically differentiate a specific neuron from another one. To be more specific, the following measurements were employed: soma surface / total surface; number of dendritic trees; number of bifurcations; number of branches; height / width; depth / height; diameter$^2 \times$ length / volume; branch order; contraction; fragmentation; partition asymmetry; Ralls ratio; bifurcation angle local; and bifurcation angle remote. Note that some of items above involve two or more of the morphometric measurements available in NeuroMorpho repository. This procedure is aimed at eliminate both the dimensionality of the measurements and size effects among different neurons. Details about such measurements are given on NeuroMorpho website.

Fig. 1 shows the mapping of the human pyramidal cells on a 3-dimensional feature space via principal component analysis (PCA) projection. To eliminate magnitude effects, the 14 measurements above are standardized prior to the PCA projection. Circles, squares and triangles depict the neurons belonging to Allman, Jacobs and Lewis datasets. PCA reduces high-correlated information on the measurements between individuals by projecting the data such that the direction of maximum dispersion is aligned

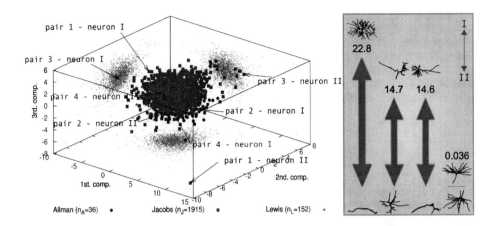

Fig. 1. Left: 3-dimensional feature-space of neuron morphometric measurements, obtained through principal component analysis (PCA) projection. Right: 4 selected pairs for analysis, selected according to the Euclidean distance on PCA feature space (given on the plot). Pairs 1 to 4 are depicted from the left to the right. The neurons of each pair are labeled so that type I cells are more complex, leading to better connectivity when compared of cells of type II. Correspondent literature: see [1] for neurons 1/I and 4/II; [12] for 1/II, 3/II and 4/I; [18] for 2/I; and [11] for cells 2/II and 3/I.

with the first component axis, and so on. Thus the 3-dimensional projection provides an objective tool for selecting "similar" or "distinct" cells according to the shape. On the right hand of the plot, 4 pairs of cells selected for analysis according to their Euclidean distance on PCA feature space are given. The greater the distance, the more the neurons are morphologically different. The three first pairs are the most distant ones adopting this criterion, while the forth one is an example of more "similar" cells.

3   Basic Concepts and Network Construction

Any *network* can be represented as a set of n *vertices* or *nodes*, interacting through a set of ℓ *connections* or *edges*. As in this work we deal with reciprocal (two-way) interactions, the presence of an edge (i, j) from i to j, $1 \leq i < j \leq n$ implicitly implies the existence of another edge (j, i) on the inverse way. Thus $\ell \leq \ell_{max} = n(n-1)/2$. A simple estimate of the level of interconnectivity on the network is the *edge density*, given by

$$\rho = 2\frac{\ell}{n(n-1)}, \tag{1}$$

i.e., $\rho = \ell/\ell_{max} \in [0,1]$. The amount $\langle d \rangle = 2\ell/n$ is the *average degree* of the network, indicating how many connections each node establishes on average with its counterparts. Denoting by $d_i, i = 1, \ldots, n$ the degree of each individual, then $\langle d \rangle = \sum_i d_i/n$. The average degree and edge density are two of the several global measurements found in the literature [8] to characterize complex networks. Many measurements derive from the degree of individual nodes. One example is the *degree entropy* [17], given by

$$h = -\sum_{d=0}^{n-1} p(d_i = d) \log p(d_i = d).$$

(2)

This measurement completes the average degree information, providing a measurement of the *dispersion* of the connectivities among the nodes.

Another measurement derived from the vertex degree is the *assortative mixing* [16]

$$r = \frac{\ell^{-1}\sum_{(i,j)} d_i d_j - \left[\ell^{-1}\sum_{(i,j)}\left(\frac{d_i+d_j}{2}\right)\right]^2}{\ell^{-1}\sum_{(i,j)}\left(\frac{d_i^2+d_j^2}{2}\right) - \left[\ell^{-1}\sum_{(i,j)}\left(\frac{d_i+d_j}{2}\right)\right]^2},$$

(3)

where $r \in [-1, 1]$. If $r > 0$, a network is said *assortative*, and vertices with higher degree tend to be interconnected, the opposite behavior observed for $r < 0$ (*disassortative network*). If $r = 0$ the vertex degrees are uncorrelated, an effect expected in randomly generated structures. It has been observed that social networks are assortative, while technological and biological networks, such as the neural network of the nematode *C. elegans* [20], for example, are disassortative [16].

Besides the degree and derived measurements, it also important to estimate the *clustering coefficient* [19] of the network,

$$c = \frac{1}{n}\sum_i \left[\frac{2\ell_i}{d_i(d_i - 1)}\right],$$

(4)

where $\ell_i \leq d_i(d_i - 1)/2$ is the number of edges shared by the immediate neighbors of node i and itself. As in the case of edge density, $c \in [0, 1]$. The clustering coefficient is associated with *small-world paradigm* [19]: clustering tends to shorten the geodesic distance (i.e. the shortest number of edges) between the nodes, even for very large systems. Brain networks are a particularly nice example of small-world structures [5], where the paradigm is related with to signal processing [14], synchrony and modularity [21].

As mentioned above, high clustering implies decrease in the average *shortest path length* between any pair of vertices on the network. This measure can be obtained by estimating the topological distance d_{ij}, i.e. the minimum number of edges separating two distinct vertices, and computing the final average. Such measurement is sensible to the number of *connected components* (disjoint subsets of interconnected vertices) that form the network: if i and j lie on distinct components, they are mutually unreachable and, by convention, $d_{ij} = \infty$. Such occurrences are generally ignored during the computation of the average shortest path length, which may lead to an improper characterization. This fact is circumvented by considering the average of the *inverse* of the shortest path lengths, namely the *global efficiency* [15]:

$$E = \frac{1}{n(n-1)}\sum_i \sum_{j \neq i} \frac{1}{d_{ij}}.$$

(5)

$E \in [0, 1]$, attaining its maximum value for complete networks, i.e. when all the vertices are interconnected and $\rho = 1$.

In this work the presented measurements are employed to characterize the neuronal networks generated as follows.

As mentioned in Sect. 2, digital reconstructions were selected in pairs, according to the separation among neurons of the same pair on PCA projection. Four of the pairs selected were given on Fig. 1.

For each neuron, the reconstruction matrix – with *compartment* information, see Sect. 2 for details – was read and the data were converted to a $\{X,Y,Z\}$ representation, i.e. a matrix $N \times 3$ was generated, defining the coordinates of N points or voxels, such that the origin $\{0,0,0\}$ gives the central coordinate of the soma. Moreover, information about the compartment type – soma, axon, dendrite – was omitted, the origin being defined as described only for reference-point proposes. In order to eliminate size effects of the converted reconstructions, the neuronal forms were scaled as to have the same *volume*, given in terms of the number of voxels N. The value $N = 2000$ was adopted.

The connectivity model is defined as follows: for each case (pair), n cells were placed at uniformly-distributed random positions on a 3-dimensional cube of fixed side $L = 350$. Before placement, each neuron is rotated along the three axes, according to angles also defined at random, following an uniform distribution. np of these cells are reproductions of the reconstruction type I and $n(1-p)$ are of type II, for $p = 0, 1/4, 1/2, 3/4, 1$. Several realizations of each case were carried out, with n varying from 10 to 500 at interval 10.

A connection between two given hosts on the network is defined whenever at voxels belonging to the two individuals overlap each other, i.e. they lie at the same position in the 3-dimensional cube. Figure 2 illustrates a realization involving pair 2 for $n = 100$ and $p = 1/2$, displaying several connections established as described.

(a) (b)

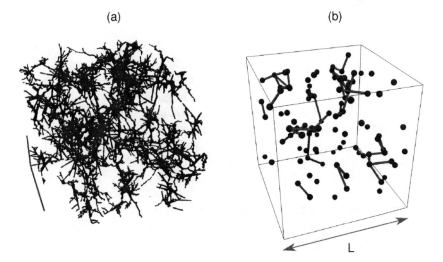

Fig. 2. Network model: (a) an ensemble of neurons of type I and II for $p = 1/2$. (b) The correspondent network defined by the connectivity of such neurons, as described in the text. The neurons populate a cubic region of fixed side L. The positions of equivalent nodes in (b) correspond to the reference-point of neuronal reconstructions – see text for details.

4 Ising Ferromagnetic Dynamics

Besides the topology, another important aspect of a complex system are the *dynamical processes* it may undergo. It is straightforward to address such dynamics by associating either a single or an array of characteristics to each individual element of the system and then allow such features to evolve along time. As usually the attributes of a given host are strongly – if not directly – correlated with the attributes of its immediate neighbors, the topology itself plays an important role on the dynamics evolving. In this paper, we promote the simplest dynamics over the network, by associating a single and binary characteristic to any of its hosts. At a given time t, the network is under a given *state*

$$\mu = \{\mu_1, \mu_2, \ldots, \mu_n\}, \tag{6}$$

where μ_i is the value associated to the i-th host. Note that since μ_i assumes either one of only two possible values, then 2^n distinct states are possible, a high amount even for small values of n. Though it is expected that any given state μ occurs with probability $p_\mu = 2^{-n}$, in real world certain configurations predominate, and diverse models in the literature deal with this fact. Certainly the most known and studied model of Statistical Mechanics, the Ising model is an example. In this model, if the system is under thermal equilibrium and at a specific temperature T, the probability for a given state follows the *Boltzmann distribution*

$$p_\mu = \frac{1}{Z} e^{-\beta E_\mu}, \tag{7}$$

where Z is a normalizing factor, $\beta = T^{-1}$ is the inverse of temperature (actually given in terms of energy), and

$$E_\mu = - \sum_{(i,j)} \mu_i \mu_j \tag{8}$$

is the *energy of the state* μ, where $\mu_i = \pm 1$, $i = 1, \ldots, n$ are the *spins* associated with the hosts. The sum is evaluated at the edges of the network, indicating the dependence of the interactions between connected vertices. The equation above in fact gives a particular case of the internal energy at each state, where no external fields are present, and the system is under *ferromagnetic regime*. Another simplification is to consider the *interaction parameter* J_{ij} that defines the strength of magnetic interaction at the nearest-neighbors i, j as being constant and equal the unity for all the edges, so this quantity vanishes on the sum above. Note also from eq. (7) that states where the energy is lower are more probable: arrangements where all the *spins* μ_i are either *up-* (+1) or *down-aligned* (-1), are favored. In such conditions, the system is said to undergo a *spontaneous magnetization*, characterized by $|m| = 1$, where

$$|m| = \frac{1}{n} \left| \sum_i \mu_i \right| \tag{9}$$

is the *magnitude of the magnetization per spin*. As the system is heated by the increase of the temperature T, the alignment of the spins tends to be uncorrelated, and the magnetization vanishes.

In this paper, the Ising model is applied only to the giant component of the neuronal networks, when $n = 500$, in order to avoid influences of small clusters and isolated hosts on the overall dynamics. Note that in this case, we replace n by n_g in the eqs. (6) and (9), where n_g is the number of hosts belonging to the giant component.

5 Results and Discussion

Figure 3 presents the behavior of some important global network measurements considering the 4 cases of study presented, with the variation of parameter p. The effect of the neuronal shape is evident, remarkably for pair 1. Indeed, for this case, both edge density and clustering coefficient seem to be linearly dependent on the parameter p (second and third plots at the leftmost column). It is interesting to observe how similar are cases 2 and 4. These cases differ at most for clustering coefficient c and global efficiency E. As the increase on the clustering coefficient shorten paths along the network [19], the fact that the pair with higher increase in clustering coefficient present lower efficiency, and vice-versa, among pairs 2 and 4, indicate the the presence of another factor(s) governing the network topology, neither captured by the remaining measurements, nor by the choice methodology adopted (Sec. 2), where pair 4 is viewed as a case of similar cells, in opposition to pair 2.

We can observe in Fig. 4 the effect of host morphology over zero-field Ising ferromagnetic dynamics, through the curves of decay of the magnetization – top row – as a function of the temperature of the system ($T = 0.2, 0.4, \ldots, 5$). Again, the curves are more sensible to p for pair 1 (leftmost plots). The energy per spin $e = E_\mu/n_g$ and the average spin cluster size as a fraction of number of hosts on giant component are also depicted, given at second and third rows, respectively. Again, as in the case of topological measurements, cases 2 and 4 feature similar behavior. Moreover, the more attenuate decay of magnetization for higher values of p – more developed hosts – suggests that synchronous behavior may be favored by the complexity of the dendritic arborization, as the spontaneous magnetization remains for higher temperatures.

Table 1 presents some interesting correlations between topological measurements and Ising variables, considering the 4 pairs discussed as well as three other cases, omitted here for the sake of clarity. The association between high network density and high connectivity entropy seems to retard the drop of magnetization, in agreement with the hypothesis risen on previous paragraph.

Table 1. Correlation coefficient r_c between topological measurements and dynamical variables

r_c	$e(T = 0.2)$	$T_e{}^b$	$T_m{}^c$	$\langle n_s/n_g \rangle^d$
E^a	-0.998	0.964	0.922	0.976
ρ^a	-0.886	0.946	0.962	0.853
h^a	-0.955	0.952	0.923	0.901

[a]for $n = 500$; [b]value of T observed where the increase rate of e is maximum; [c]value of T observed for the maximum decrease rate of $|m|$; [d]average of mean spin-cluster size over the employed range of T.

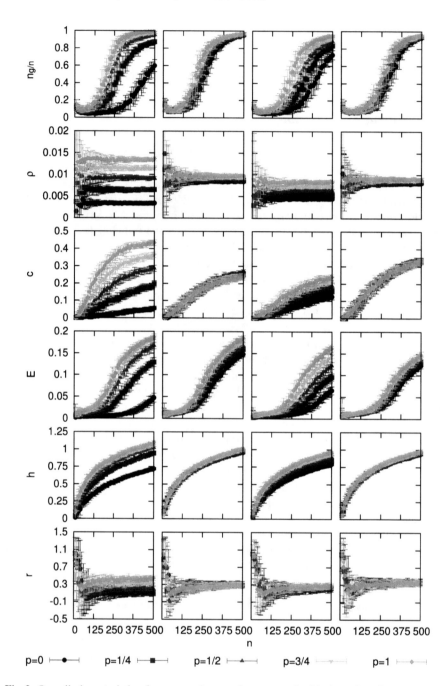

Fig. 3. Overall characteristics for neuronal networks generated with the pairs of neurons given on figure 1. From left to right: pairs 1 to 4. From the first to last row: ratio of giant component size over network size n_g/n, edge density ρ, clustering coefficient c, global efficiency E, degree entropy h and assortative mixing r. See text for details.

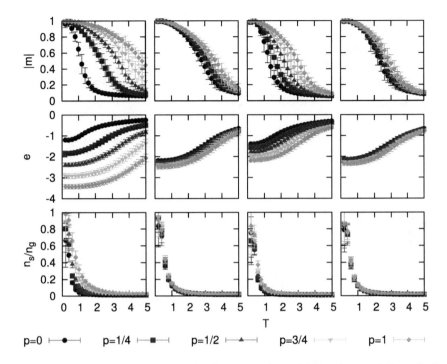

Fig. 4. Thermodynamical properties of Ising ferromagnetic model for pairs 1 to 4. From the first to last row: magnitude of magnetization per spin $|m|$, internal energy per spin e and fraction of giant cluster of spins to the giant component size, where the model is applied. See text for details.

6   Concluding Remarks

In the present paper we addressed the effect of the host morphology over both the global characteristics of a complex network and actuating dynamic processes it undergoes, by adopting realistic and 3-dimensional digital reconstructions of human pyramidal neurons, and Ising dynamics. This effect was explored by the variation of the ratio p of one neuron to its counterpart, for different pairs of neurons selected. As the criterion of choosing "different" neurons to represent a pair for the generated systems is based only on a 3-dimensional projection of the individual morphological measurements, other factors certainly are present, which explain the similarity found for pairs 2 and 4, for example, as well as the different behavior among cases 2 and 3. However, the results presented so far show that the impact of the individual morphology over global aspects of the neuronal systems is clear, specially considering cases 1 and 3. Nonetheless, the aspect of shape over the connectivity and dynamic behavior of spatial networked systems has received little attention along the last years. In networks such as the neuronal ones, the shape of individual constituents is of fundamental importance. We believe that the results so far illustrated pave the way for new perspectives on the influence of individual aspects over global system behavior and function.

Acknowledgement. The authors thank S. N. Taraskin for helpful comments and suggestions. R. A. P. Silva, M. P. Viana and L. da F. Costa are grateful to financial support by FAPESP (grants 07/54742-7, 07/50882-9 and 05/00587-5, respectively). Costa is also grateful to CNPq (grants 301303/06-1 and 573583/08-0).

References

1. Anderson, K., Bones, B., Robinson, B., Hass, C., Lee, H., Ford, K., Roberts, T.A., Jacobs, B.: The morphology of supragranular pyramidal neurons in the human insular cortex: a quantitative Golgi study. Cereb. Cortex 19(9), 2131–2144 (2009)
2. Ascoli, G.A.: Mobilizing the base of neuroscience data: the case of neuronal morphologies. Nat. Rev. Neurosci. 7, 318–324 (2006)
3. Ascoli, G.A., Scorcioni, R.: Neuron and Network Modeling. In: Zaborszky, L., Wouterlood, F.G., Lanciego, J.L. (eds.) Neuroanatomical Tract-Tracing, vol. 3, pp. 604–630. Springer, New York (2006)
4. Boccaletti, S., Latora, V., Moreno, Y., Chavez, M., Hwang, D.-U.: Complex networks: structure and dynamics. Phys. Rep. 424, 175–308 (2006)
5. Bullmore, E., Sporns, O.: Complex brain networks: graph theoretical analysis of structural and functional systems. Nat. Rev. Neurosci. 10(3), 186–198 (2009)
6. da Costa, L.F., Manoel, E.T.M.: A percolation approach to neural morphometry and connectivity. Neuroinform. 1 (1), 65–80 (2003)
7. da Costa, L.F., Coelho, R.C.: Growth-driven percolations: the dynamics of connectivity in neuronal systems. Eur. Phys. J. B 47, 571–581 (2005)
8. da Costa, L.F., Rodrigues, F.A., Travieso, G., Boas, P.R.V.: Characterization of complex networks: a survey of measurements. Adv. Phys. 56 (1), 167–242 (2007)
9. Eberhard, J.P., Wanner, A., Wittum, G.: NeuGen: A tool for the generation of realistic morphology of cortical neurons and neuronal networks in 3D. Neurocomputing 70(1-3), 327–342 (2006)
10. Gleeson, P., Steuber, V., Silver, R.: Neuroconstruct: a tool for modeling networks of neurons in 3D space. Neuron. 54, 219–235 (2007)
11. Hayes, T.L., Lewis, D.A.: Magnopyramidal neurons in the anterior motor speech region. Dendritic features and interhemispheric comparisons. Arch. Neurol. 53(12), 1277–1283 (1996)
12. Jacobs, B., Schall, M., Prather, M., Kapler, E., Driscoll, L., Baca, S., Jacobs, J., Ford, K., Wainwright, M., Treml, M.: Regional dendritic and spine variation in human cerebral cortex: a quantitative Golgi study. Cereb. Cortex 11(6), 558–571 (2001)
13. Koene, R.A., Tijms, B., van Hees, P., Postma, F., Ridder, A., Ramakers, G.J.A., van Pelt, J., van Ooyen, A.: NETMORPH: A framework for the stochastic generation of large scale neuronal networks with realistic neuron morphologies. Neuroinform. 7, 195–210 (2009)
14. Lago-Fernández, L.F., Huerta, R., Corbacho, F., Sigüenza, J.A.: Fast response and temporal coherent oscillations in small-world networks. Phys. Rev. Lett. 84, 2758–2761 (2000)
15. Latora, V., Marchiori, M.: Efficient behavior of small-world networks. Phys. Rev. Lett. 87, 198701 (2001)
16. Newman, M.E.J.: Assortative mixing in networks. Phys. Rev. Lett. 89, 208701 (2002)
17. Wang, B., Tang, H., Guo, C., Xiu, Z.: Entropy optimization of scale-free networks' robustness to random failures. Phys. A 363(2), 591–596 (2005)
18. Watson, K.K., Jones, T.K., Allman, J.M.: Dendritic architecture of the von Economo neurons. Neurosci. 141(3), 1107–1112 (2006)

19. Watts, D.J., Strogatz, S.H.: Collective dynamics of 'small-world' networks. Nature 393, 440–442 (1998)
20. White, J.G., Southgate, E., Thomson, J.N., Brenner, S.: The structure of the nervous system of the nematode *Caenorhabditis elegans*. Phil. Trans. R. Soc. Lond. B 314, 1–340 (1986)
21. Yu, S., Huang, D., Singer, W., Nikolic, D.: A small world of neuronal synchrony. Cereb. Cortex 18, 2891–2901 (2008)
22. Zubler, F., Douglas, R.: A framework for modeling the growth and development of neurons and networks. Front. Comput. Neurosci. (2009), doi: 10.3389/neuro.10.025.2009

Assessing Respondent-Driven Sampling in the Estimation of the Prevalence of Sexually Transmitted Infections (STIs) in Populations Structured in Complex Networks

Elizabeth M. Albuquerque[1], Cláudia T. Codeço[2], and Francisco I. Bastos[3]

[1] Oswaldo Cruz Foundation, Avenida Brasil, 4365, Biblioteca de Manguinhos suite 229,
Rio de Janeiro, RJ, Brazil
emaciel@icict.fiocruz.br
[2] Oswaldo Cruz Foundation, Avenida Brasil, 4365, PROCC, Rio de Janeiro, RJ, Brazil
codeco@fiocruz.br
[3] Oswaldo Cruz Foundation, Avenida Brasil, 4365, Biblioteca de Manguinhos suite 229,
Rio de Janeiro, RJ, Brazil
and
CAPES/Fulbright Visiting Scholar at Brown University, RI, USA
bastos@icict.fiocruz.br

Abstract. When a sampling frame for a given population cannot be defined, either because it requires expensive/time-consuming procedures or because it targets a stigmatized or illegal behavior that may compromise the identification of potential interviewees, traditional sampling methods cannot be used. Examples include "hidden populations" of special relevance for public health, such as men who have sex with men (MSM), sex workers and drug users. Since the late 1990s, a network-based method, called Respondent-Driven Sampling (RDS) has been used to assess such "hidden populations".This paper simulates data from hidden populations, in order to assess the performance of prevalence estimators in different scenarios built after different combinations of social network structures and disease spreading patterns. The simulation models were parameterized using empirical data from a previous RDS study conducted on Brazilian MSM. Overall, RDS performed well, showing it is a valid strategy to assess hidden populations. However, the proper analysis of underlying network structures and patterns of disease spread should be emphasized as a source of potential estimate biases.

1 Introduction

Several factors may compromise the accurate characterization of a population. In particular, there are situations where the defining characteristics are difficult to observe — as a consequence of prohibitive costs/complex logistics or its association with an illegal and/or stigmatized behavior. In Public Health, populations with these characteristics are named at-risk hidden or hard-to-reach populations [5]. In the context of Sexually Transmitted Infections (STIs), some important examples are: men who have sex with men (MSM), sex workers (SW), and people who misuse drugs. When a sampling frame cannot be defined, alternative sampling strategies are necessary and one popular approach is the "snowball" sampling, which uses the visible part of the population and

L. da F. Costa et al. (Eds.): CompleNet 2010, CCIS 116, pp. 108–118, 2011.

their social networks to identify and recruit unknown members from the hidden population [11]. The chain-referral process begins with the selection of a small number of individuals of the population (the seeds), who recruit other members of the population from their personal network to form the *first wave*. The researcher then samples individuals from this first wave and ask them to enroll their contacts, to form the *second wave*. The process is repeated until the target sample size is achieved. This sampling process generates chains of successive pairs of recruiter-recruited individuals.

There is a variation of the snowball sampling called Respondent-Driven Sampling (RDS), proposed by Heckathorn [5], which has been largely used for sampling at-risk populations, estimate infections/illnesses prevalence [8] and to assess risk factors. RDS also produces referral chains but contrasting with the traditional snowball sampling, RDS provides a limited number of coupons to each participant, who is stimulated through incentives, to enroll members of their contact networks themselves through the delivery of these coupons. The incentive system guarantees the collaboration of the participants in the recruitment process. By constraining the number of coupons to 2 or 3, biases associated with massive recruiting by a single individual (i.e. a super-recruiter) are minimized and referral chains may grow longer, increasing the chance of moving farther into the social networks, as discussed in detail elsewhere [14].

Alternative weighted estimators of prevalence (i.e. the number of cases of a specific infection/disease present in a given population at a certain time) for RDS samples have been proposed [5,6,17] and their accuracy evaluated by simulation studies (as reviewed in [3]). In these studies, virtual populations structured into networks are created, from which RDS samples are taken and compared to the known population prevalence. Overall, these studies suggest that RDS prevalence estimates are not sensitive to the strategy used to choose the seeds, whether random or proportional to the individual's degree [14]. They can be biased, though, when seeds are chosen according to the infectious status (infected vs. uninfected), specially if there are clusters of infection in the network [3]. Large sample sizes seem to be important to reduce the estimates uncertainty [17], although too large samples may eventually have the opposite effect [3].

Simulation results are model specific, i.e., they are sensitive to the type of network created and should not be extrapolated to other scenarios. Many RDS studies have used unrealistically simple network topologies, most likely overestimating their accuracy [4]. Gile and Handcock [3] were the first to assess RDS estimates using a more realistic network structure. In their study, they set up several scenarios by varying the clustering of infections in the network and the degree ratio of infectious vs. non-infectious.

In the present paper we build a simulation environment based on parameters obtained from an empirical RDS study conducted in Campinas, Brazil, targeting the MSM population. The study recruited 689 subjects over one year and estimated the seroprevalence of HIV and syphilis, as well as their sociodemographic and behavioral traits [10]. In our simulation study we considered different scenarios of infection distribution: random, degree-dependent, individual covariate-dependent, and following a transmission model. RDS was implemented with both incomplete and complete recruitment, and prevalence estimates were calculated using the estimator originally defined by [6]. The results, presented as a panel of different scenarios, were generated under varied realistic conditions.

2   Methods

Our simulation procedure is presented in three steps. First, we performed an exploratory analysis of the empirical data from the Campinas MSM study [10] to identify individual factors associated with network formation and infectious status. Secondly, we launched the simulation process, with the construction of a baseline population with attributes mimicking the empirical MSM population. Different scenarios were then built, varying the distribution of "infected" vs. "non-infected" individuals in the population. Finally, we took samples of the virtual population using RDS considering two alternative recruitment processes (with a fixed and a random number of coupons) and compared the estimated prevalences as a function of network structure, infection distribution, and recruitment. Figure 1 summarizes these steps.

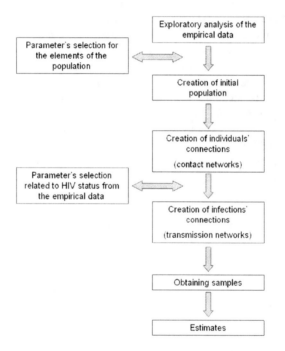

Fig. 1. Diagram of the algorithm used for the simulations

2.1   Empirical Data

The empirical data comes from a RDS study conducted between October 2005 and October 2006, targeting the MSM population from the metropolitan area of Campinas, SP, Brazil [10]. Individuals aged 15 year or more, capable of reading and signing an informed consent, who had practiced oral or anal sex with a male partner in the last 6 months were recruited. In addition to a self-applied instrument, recruitees were asked to take rapid tests for syphilis and HIV after counseling. All individuals received post-test counseling and were referred to public health units whether applicable. The study recruited 689 MSM, among whom 658 were eligible and enrolled in the study. The study used thirty "seeds".

The answer to the question: "How many MSM you know that you could get in touch with (in person or by telephone) and have met in the past month?" was used as a measure of the subject's personal network size (degree). The average degree was 21.98, varying between 0 and 700, and 95% of the recruitees reported to have up to 80 contacts. An exponential model [18] with parameter 0.08 fits well the empirical degree distribution ($R^2 = 0.59$) and was used to draw degrees for the simulation study.

Traits associated with homophily were assessed as well. Homophily may be defined as the tendency of a given individual to be linked to people who share the same characteristics with them. Contigency tables and their respective statistics were used to identify variables associated with "recruiter-recruitee" pair formation, including: education, income, race, sexual orientation (homosexual, bisexual), and age. Among them, only sexual orientation and age were found to be assortative ($p < 0.01$), with estimated probability of pairing between "homosexual" (H) and "bisexual" (B): $Pr(H - B) = 0.211$, $Pr(H - H) = 0.789$, $Pr(B - B) = 0.348$, and $Pr(B - H) = 0.652$. The relation between the age of recruiter and recruitee was well represented by the linear model $age_{recruited} = 13.58 + 0.453age_{recruiter}$ ($R^2 = 0.23$) indicating that individuals in this population tend to relate with those with similar age or slightly younger.

2.2 Simulation Scenarios

A baseline synthetic population of 25,000 individuals, varying in age, sex orientation, income, education, and degree was simulated. Individual traits (all but degree) were sampled from the Campinas' study empirical distributions, while the degree was sampled from an exponential distribution with parameter 0.8 (fitted to the empirical data). To assess the effect of network structure on RDS, four network generators were defined: a) a random connection network ($L01$); b) a sexual orientation-dependent network ($L02$); c) an age-dependent network ($L03$); and d) a sexual orientation plus age dependent network ($L04$). To generate each network, a link between a pair of individuals i and j was created with probability given by their degree and attributes (age and sex orientation), after the empirical study (Table 1). With this procedure, four differently-networked populations were created, composed by the same individuals, but connected in different ways.

Infected versus non-infected status. We simulated four alternative patterns of infection distribution in the synthetic populations, taking HIV as the putative infection (Table 1). In all scenarios, the prevalence used was 0.2, as in [3]. Recent findings from Thailand, Cambodia, and Senegal converge on ~ 0.2 for the prevalence of HIV in local populations of MSM [1], the upper limit of the pooled prevalence for MSM been in Brazil — 13.6% ($95\% \, CI : 8.2 - 20.2$) [9].

Pattern A consisted in the random distribution of infected cases in the population; In *Pattern B*, the probability of being infected is weighted by the individual degree. Therefore, individuals with more contacts are more likely to be infected. *Pattern C* uses an empirical risk model fitted to the original database. A statistical analysis, using Fisher's, Chi-square and t-Student, was carried out to identify variables associated with infection in the Campinas study. Age, socio-economic class, serological status for syphilis, education, sexual orientation and race were investigated. A logistic regression model [7] , with variables age, socio-economic strata, and education, was fitted to the data, using HIV infection status as response variable. From this model, individuals' probability of being infected (p) was obtained (Table 1). Finally, *Pattern D* was built dynamically by implementing a disease transmission process: first, 50 randomly selected individuals were infected. Then, their contacts were infected with probability 0.7%. This process was repeated until reaching the target prevalence.

Combining four population structures with four infection distribution patterns, we generated 16 different population scenarios to be analysed using RDS.

Table 1. Probability of infection in the four simulated scenarios, assuming a final prevalence of 0.2. X_1 is age, X_2, X_3 and X_4 are the socioeconomic categories and X_5 is schooling; d is degree.

Infection model	Probability of individual i be infected
A (random)	$\pi_i = 0.2$
B (dependent on degree)	$\pi_i = 0.2 \frac{d_i}{\sum d}$
C (empirical)	$log(\frac{p}{1-p}) = \bullet\, 2.28 + 0.076X_1 \bullet\, 1.195X_2 \bullet\, 1.399X_3 \bullet\, 0.68X_4 \bullet\, 0.111X_5$
D (transmission)	Dynamic transmission model (see text)

RDS Sampling. The next step consisted in the implementation of the RDS sampling process, with a target sample size of 500 subjects, starting with 5 seeds, and giving 3 coupons per individual [5,6]. Seeds were sampled weighting by individual degree, as usually done in the practice. Two recruitment processes were then considered:

Structured recruitment. All participants makes the maximum possible recruitments. In this scenario, less than three recruitees per recruiter only occurs if the recruiter's degree was less than three or if all his contacts already belonged to the sample (sampling without replacement).

Random recruitment. The number of individuals recruited per person is a random number between 0 − 3. The proportions observed in the Campinas' study were used: 47%, 24%, 18%, and 11% of the interviewees recruited 0,1,2, and 3 persons, respectively.

2.3   Estimation

The accuracy and precision of two prevalence estimates were assessed using the developed simulator. First, the *simple estimate* is the proportion of individuals that were infected in the sample. This estimate was used in order to compare it with the *Heckathorn's estimate* [6], that takes in consideration the average degree of the subjects in each group (infected vs. non-infected) as well as the number of links between infected and non-infected individuals.

To calculate the estimate's precision, the sampling process was repeated 100 times for each of the 16 scenarios described above, 50 of them using structured recruitment and the other 50, using random recruitment. The simulation and sampling algorithms (available under request) were implemented in R 2.9.1 [16], and it is easily adaptable to any modality of snowball sampling, as well as for the simulation of populations with different characteristics.

3   Results

Figure 2 illustrates RDS chains generated by the simulation. Each circle represents an individual in the sample and the bigger ones represent "seeds". The complete recruitment process produced five homogeneous and relatively short chains. The random recruitment, on the other hand, produced heterogeneous chains with some branches stretching through several waves while others were fruitless or ended after one or two waves. The number of waves varied from 0 up 93. This heterogeneity is similar to the observed pattern in the Campinas study. In practice, fruitless chains may be turned into fruitful ones by the inclusion of new seeds or stimulating interviewees to seek additional recruitees. However, such hypothetical interventions were not explored here.

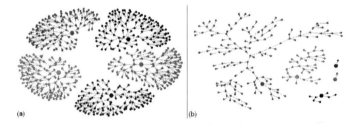

(a)   (b)

Fig. 2. Examples of a sampled network obtained with complete recruitment (a) and random recruitment (b)

3.1   RDS with Structured Recruitment

Table 2 and Figure 3 show that the simple proportion estimate is biased in scenarios where infection is correlated with the individual degree and, to a lower extent, in the scenario with clustered infections (transmission model). In scenarios where infection is distributed randomly or according to individual attributes, the simple proportion estimate was found to be unbiased. Heckathorn's estimator successfully corrected the estimated prevalence for degree-dependent infection distributions and clustered infection distributions generated by transmission. In some cases, though, when both methods yielded unbiased estimates, Heckathorn's estimator tended to be less precise than the simple one.

When samples were obtained by random recruitment, both estimators were found to be inaccurate (Table 2 and Figure 4), and some samples presented highly discrepant prevalence estimates (prevalence = 1, for example, in scenario $L01$-B). Note that the y-axis scale in these boxplots are twice as large as those depicted in Figure 3 (structured recruitment). This result suggests that estimate uncertainty decreases when all subjects managed to recruit the number of recruitees they are expected to. Accuracy was compromised in other scenarios as well. While in the structured recruitment, the Heckathorn

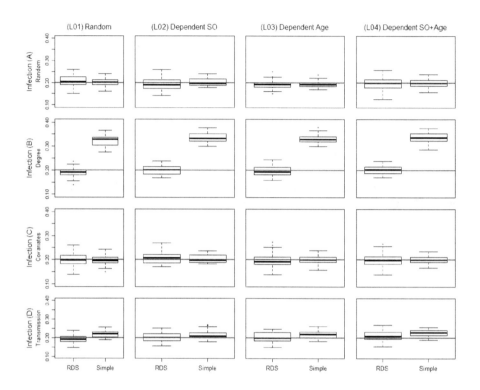

Fig. 3. Distribution of prevalence estimates obtained with the structured recruitment

estimates never exceeded 10% error, under the random recruitment, it reached 23% error in scenario *A-L04*, 32% error in scenario *B-L03*, and 15% in scenario *C-L03*. These are scenarios that involved age-dependent pair formation. However, in other scenarios where the simple estimation behaved poorly (*Infection B*), Heckathorn's estimator seems to be unbiased.

4   Discussion

RDS is a sampling methodology with two major aims: One, to identify members of hidden populations and two, to provide a probability sample for the estimation of population properties. Regarding the first aim, design questions should be considered involving the number and type of seeds and number of coupons per recruiter. Previous studies have shown that choosing seeds according to their infection status can bias the estimates [3] as, in most cases, infection status is a source of homophily in real networks. Here, we simulated scenarios with a fixed number of seeds which were chosen randomly according to their degree distribution. No effect of such specific choice on estimates accuracy is reported by other simulation studies [14]. The optimal number of coupons distributed per recruiter has also been a matter of debate. Here, we analyzed the effect of incomplete recruitment on the homogeneity of the chain referrals and the

Table 2. Median estimated prevalence calculated from the RDS samples (by simple and Heckathorn's RDS estimators) and p-values of Wilcoxon test for median comparison

Infection	Network	Complete			Random		
		Simple	RDS	p-value	Simple	RDS	p-value
A	L01	0.204	0.205	0.48	0.200	0.197	0.64
	L02	0.197	0.190	0.06	0.183	0.195	0.46
	L03	0.193	0.192	0.41	0.210	0.173	0.18
	L04	0.198	0.198	0.62	0.182	0.153	0.04
B	L01	0.329	0.192	< 0.01	0.356	0.208	< 0.01
	L02	0.333	0.201	< 0.01	0.333	0.195	< 0.01
	L03	0.329	0.194	< 0.01	0.379	0.265	< 0.01
	L04	0.338	0.202	< 0.01	0.348	0.201	< 0.01
C	L01	0.196	0.198	0.73	0.188	0.210	0.47
	L02	0.199	0.208	0.48	0.196	0.190	0.67
	L03	0.200	0.193	0.43	0.212	0.229	0.29
	L04	0.201	0.199	0.64	0.203	0.187	0.32
D	L01	0.222	0.192	< 0.01	0.219	0.172	0.02
	L02	0.210	0.202	< 0.01	0.202	0.181	0.17
	L03	0.219	0.200	< 0.01	0.221	0.200	0.32
	L04	0.230	0.209	< 0.01	0.211	0.171	0.04

accuracy of the estimates. Our findings show that, under complete recruitment, the target sample size is achieved rapidly, resulting in short chains with a small number of waves. The sample grows at a geometric rate [6] and the resulting chains are homogenous in structure and size. Fast achievement of target sample size can be a problem. The shorter the chain, the closer its members will be to the seed. Since the ultimate goal of RDS is to obtain a sample that is independent of the seed, to allow it to grow longer is a basic requirement. In combination, these results suggest that a pre-specified (not small) number of waves is a better stopping rule than a given sample size. The more realistic incomplete recruitment, on the other hand, produced more heterogeneous chains, where the estimates were less precise and, in some scenarios, even biased. These results indicate that attention should be given to ensure the completeness/homogeneity of the recruitment process. In practice, this is a challenge, as different populations, such as MSM, drug users, populations of high and low income countries, etc., behave differently in RDS studies [8]. It is important to highlight the negative impact of incomplete recruitment on the accuracy of both estimators. This is relevant since, in real-life situations, each empirical study builds on a single sample rather than on several samples, as simulated. When a complete recruitment takes place, estimates are more precise.

Salganik [15], using different simulation procedures, obtained comparable results, with the estimates converging to the true population value. He also found the variability of the RDS estimates to be higher than those obtained with simple random sampling. Simulation studies [4,14] indicate that the larger the size of the sample, the closer to the true values are the estimates. However, in [14], the author shows that the estimates invariably overestimates the actual prevalence. As shown by our study, estimates can either over or underestimate actual prevalences. Also noteworthy is that the differences

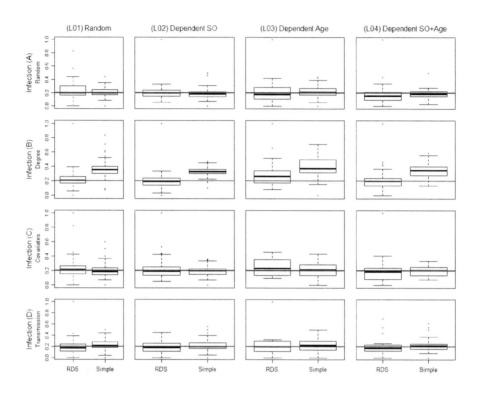

Fig. 4. Distribution of prevalence estimates obtained with the random recruitment

between the estimates and the true value of the parameter were modest in [14], in the order of the third decimal digit, which was not corroborated by our study. One of the reasons for this discrepancy between [14] and our findings may be due to the fact their simulation used sampling with replacement, while the simulations showed in this paper were generated by processes without replacement.

Regarding the nature of each infection pattern, for infections where the size of the contact networks influenced the exposure to infection (scenarios with infection pattern B), the weighting estimate proposed by Heckathorn showed good results, whereas the simple model tended to overestimated actual prevalences. Weighting yielded more accurate results in situations where infections are dependent to personal contact as well (*infection D*). Both cases mentioned above seem to correspond to situations found among at-risk populations, since it is expected individuals who know more people to be especially vulnerable to infections that are spread through intimate physical interaction [13]. Different estimation proposals for weighting have being developed [17,12], and require further analyses using additional empirical data and simulations.

Although our findings speak in favor of the use of RDS, especially with structured recruitment, caution is mandatory since additional sources of variability and imprecision may exist [3]. When comparing estimates using [6] and [17] estimators, the authors verify the latter was more accurate. RDS is currently a popular approach for the estimation of the prevalence of HIV and other STIs, but there are only a few studies that address the accuracy of those estimates. A cautionary note on the interpretation of these estimates and in their use in the formulation and monitoring of public policies was evidenced by the present study and was recently discussed by [4]. Simulation studies that use parameters obtained from empirical studies carried out in different contexts and/or different populations are key to verify the feasibility of monitoring epidemic trends over time using RDS methodology. Social networks are dynamic entities by their very definition, which reshape themselves over the course of an epidemic [2]. Every effort should be made in order to distinguish the influence of different factors on the accuracy of estimates. More refined models should take into account not only the intrinsic variability of the social networks structure, infection patterns, and estimation methods, but also the variability of the measured parameters over time.

Acknowledgments. We thank Maeve Mello for our helful discussions about the Campinas' RDS study and CAPES/Fullbright, FAPERJ and CNPq for their support with career grants. This work was also supported by CNPq (Edital Universal).

References

1. Baral, S., Sifakis, F., Cleghorn, F., Beyrer, C.: Elevated risk for hiv infection among men who have sex with men in low- and middle-income countries 2000-2006: a systematic review. PLoS Med. 4(12), e339 (2007),
   http://dx.doi.org/10.1371/journal.pmed.0040339,
   doi:10.1371/journal.pmed.0040339
2. Boily, M.C., Bastos, F.I., Desai, K., Chesson, H., Aral, S.: Increasing prevalence of male homosexual partnerships and practices in britain 1990-2000: but why? AIDS 19(3), 352–354, author reply 354–5 (2005)
3. Gile, K.J., Handcock, M.S.: Respondent-driven sampling:: an assessment of current methodology. Sociological Methodology, 1467–9531 (2010)
4. Goel, S., Salganik, M.J.: Assessing respondent-driven sampling. Proc. Natl. Acad. Sci. USA 107(15), 6743–6747 (2010), http://dx.doi.org/10.1073/pnas.1000261107, doi:10.1073/pnas.1000261107
5. Heckathorn, D.: Respondent-driven sampling: a new approach to the study of hidden populations. Social Problems 44, 174–199 (1997)
6. Heckathorn, D.: Respondent-driven sampling ii: deriving valid population estimates from chain-referral samples of hideen populations. Social Problems 49, 11–34 (2002)
7. Hosmer, D., Lemeshow, S.: Applied Logistic Regression. Jonh Wiley & Sons, Chichester (2000)
8. Malekinejad, M., Johnston, L.G., Kendall, C., Kerr, L.R.F.S., Rifkin, M.R., Rutherford, G.W.: Using respondent-driven sampling methodology for hiv biological and behavioral surveillance in international settings: a systematic review. AIDS Behav. 12(4 Suppl.), S105–S130 (2008), http://dx.doi.org/10.1007/s10461-008-9421-1
9. Malta, M., Magnanini, M.M.F., Mello, M.B., Pascom, A.R.P., Linhares, Y., Bastos, F.I.: Hiv prevalence among female sex workers, drug users and men who have sex with men in brazil: a systematic review and meta-analysis. BMC Public Health 10, 317 (2010), http://dx.doi.org/10.1186/1471-2458-10-317, doi:10.1186/1471-2458-10-317
10. Mello, M., A, A.P., M, M.C., Tun, W., Júnior, A.B., Ilário, M., Reis, P., Salles, R., Westman, S., Díaz: Assessment of risk factors for hiv infection among men who have sex with men in the metropolitan area of campinas city, brazil, using respondent-driven sampling. Tech. rep., Population Council (2008)

11. Morris, M.: Network Epidemiology: A handbook for survey design and data collection. Okford University (2004)
12. Poon, A.F.Y., Brouwer, K.C., Strathdee, S.A., Firestone-Cruz, M., Lozada, R.M., Pond, S.L.K., Heckathorn, D.D., Frost, S.D.W.: Parsing social network survey data from hidden populations using stochastic context-free grammars. PLoS One 4(9), e6777 (2009), http://dx.doi.org/10.1371/journal.pone.0006777, doi:10.1371/journal.pone.0006777
13. Romano, C.M., de Carvalho-Mello, I.M.V.G., Jamal, L.F., de Melo, F.L., Iamarino, A., Motoki, M., Pinho, J.R.R., Holmes, E.C., de Andrade Zanotto, P.M., V. G. D. N. Consortium: Social networks shape the transmission dynamics of hepatitis c virus. PLoS One 5(6), e11,170 (2010), http://dx.doi.org/10.1371/journal.pone.0011170, doi:10.1371/journal.pone.0011170
14. Salganik, M., Heckathorn, D.: Sampling and estimation in hidden populations using respondent-driven sampling. Sociological Methodology 34, 193–239 (2004)
15. Salganik, M.J.: Variance estimation, design effects, and sample size calculations for respondent-driven sampling. J. Urban Health 83(6 suppl), 98–112 (2006), http://dx.doi.org/10.1007/s11524-006-9106-x, doi:10.1007/s11524-006-9106-x
16. R Development Core Team: R: A language and environment for statistical computing. R Foundation for Statistical Computing (2008)
17. Volz, E., Heckathorn, D.: Probability based estimation theory for respondent-driven sampling. Journal of Official Statistics 24, 79–97 (2008)
18. Strogatz, S.H.: Exploring complex networks. Nature 410(6825), 268–276 (2001), http://dx.doi.org/10.1038/35065725, doi:10.1038/35065725.

Using Networks to Understand the Dynamics of Software Development

Christopher Roach[1] and Ronaldo Menezes[2]

[1] Apple, Inc., 1 Infinite Loop, Cupertino, CA 95014, United States
croach@apple.com
[2] Florida Institute of Technology, 150 W University Blvd., Melbourne, FL 32901, United States
rmenezes@cs.fit.edu

Abstract. Software engineering, being a relatively new field, has struggled to find ways of gauging the success/failure of development projects. The ability to determine which developers are most crucial to the success of a project, which areas in the project contain the most risk, etc. has remained elusive, thus far. Metrics such as SLOC (Source Lines of Code) continue to be used to determine the efficacy of individual developers on a project despite many well-documented deficiencies of this approach. In this work, we propose a new way to look at software development using network science. We examine one large open-source software development project—the Python programming language—using networks to explain and understand the dynamics of the software development process. Past works have focused on the open source community as a whole and the relationships between the members within. This work differs in that it looks at a single project and studies the relationships between the developers using the source code they create or work on. We begin our analysis with a description of the basic characteristics of the networks used in this project. We follow with the main contribution of this work which is to examine the importance of the developer within their organization based on their centrality measures in networks such as degree, betweenness, and closeness.

1 Introduction

The discipline of software engineering has been around for over 40 years; the first formal mention of the term can be traced back to a 1968 NATO publication [1] whose main purpose was to provoke thoughtful discussion on the looming "software crisis" of the time. Ever since software engineering established itself as a major field, and the production of software became a commercial activity, there has been a need for evaluating individual software developer's performance. In software companies, the problem of evaluating performance is slightly more complicated than in traditional industries because the product being produced is less tangible. For instance, in a company assembling watches, the employee can be easily evaluated by the number of watches he assembles per unit of time. In software companies, however, components produced per unit of time is impossible to judge since, in most cases, we have one piece of software being produced. This is not to say that there are no ways to do it, quite the contrary, the very first software metric, Source Lines of Code (SLOC) count, was first put into

L. da F. Costa et al. (Eds.): CompleNet 2010, CCIS 116, pp. 119–129, 2011.

use at the very end of the 1960s [2]. Thus, it true to say that for as long as there has been a discipline called software engineering, there have been metrics to quantify the importance and productivity of its practitioners.

Over forty years have passed since static, individual-based metrics such as SLOC count were first used and yet the use of these metrics to evaluate individual developers' worth within an organization still pervades the industry. However, software development projects are no longer small with few developers; most are complex, involving, in many cases, hundreds of developers and millions of lines of code. Despite the fact that it is now common knowledge that traditional metrics are poor indicators of a developer's value in today's world, they can still be found in every corner of the industry [3]. In this study, we argue that individual-based, static statistics such as SLOC count, and other similar size-based metrics, are poor indicators of a developer's overall worth within their organization. The major source of problems with these traditional metrics lies within their focus on the individual to the point of excluding the surrounding environment in which the individual does his work. We argue that a network-centric approach to measuring a developer's importance provides a much more realistic means to evaluate individual developers since it takes into account the effect the organization has on the individual developer's productivity and the effect the individual has on the organization in return.

The remainder of this report is organized as follows: In Section 2 we quickly discuss the traditional methods used in software development to evaluate developers; we also review the literature on both the use of network science to study software development and on some of the alternatives that have been prescribed to remedy the known issues with traditional metrics. Section 3 will argue for a network-centric approach because it captures: the importance of a developer to the cohesion of the team, the role of the developer in the maintenance of a stable development group, and the knowledge held by a developer and the effect of his departure to the group as a whole. We follow with our argument for using network-centric approaches. We finish the section with a description of the network measurements used in this paper. Section 4 describes the details of this study and justifies the use of the Python open-source project as well as the results of using network-centric approaches to evaluate developers. Finally, in Section 5, we discuss areas for future work and give our concluding remarks.

2   Related Works

2.1   Evaluating Software Developers

One of the common problems software development managers face in their daily activities relates to the process of evaluating developers (programmers, testers, etc.). While it would be relatively easy to quantify the amount of work done by a developer, it is quite complicated to qualify the same work. Anyone familiar with software development understands that some solutions to problems may be harder to code than others. Hence the amount of work produced may not reflect the effort a developer has put into the process.

There have been many approaches used to evaluate software developers:

Source Lines of Code: SLOC is likely to have been the first metric to evaluate developers. In a process of software development the main product is a program (a piece of software). Thus, it is natural to think that developers could be evaluated by the number of lines of code they contribute to the process. Clearly this is not a good approach since the number of lines does not imply quality, and worse, the complexity of the lines of code produced is disregarded in simple counts.

Number of Commits: It is common for companies to keep track of the number of modifications a developer makes to the project. The idea is that more prolific developers tend to make more modifications (commits). However this is highly questionable, since not all commits have the same importance.

Number of Issues Found: This measure is very similar to the number of commits, but closely related to issues of bugs found in the development. Similar to the cases above, this is not a good indication because different bugs have different importance. In fact, the location (file) where the bug is found is also important.

Hours worked (extra hours): Many companies evaluate developers by their level of dedication to the project. The number of hours worked on a particular project is often used as an indicator of dedication. Number of hours is a good indicator but in isolation provide little information about the effectiveness of the developer.

Innovations Introduced by the Developer: Companies today encourage developers to introduce new ideas and even reward them for introducing ideas and projects. To our knowledge this has not been formally been used as an evaluation mechanism because very few job descriptions formally require it.

Because of problems present in the approaches above, managers generally resort to less precise evaluation mechanisms such as meetings to discuss the worth of each developer from their point of view—a common approach during the evaluation cycle is to have managers discuss each person individually with some or all of the criteria mentioned above considered. While these approaches are useful and bring to the table the important perceptions of managers, the whole process is generally unfocused due to the lack of a precise reliable metric to drive the discussion.

2.2 Development Trackers

The literature also has many examples of software tools aimed at helping track the process of software development as well as the developers working on the project. Some of the most common examples of these programs include SeeSoft, CVSScan and Chronia.

SeeSoft [4] is essentially a versioning tool for visualizing software development at the source-code line level. Modifications in the code (lines) can be associated with any statistic of interest such as: most-recently changed, least-recently changed, number of changes, etc. Although not proposed for evaluation of developers it could be easily adapted to be used to track the number of lines changed by developers and hence automate metrics such as SLOC.

CVSScan [5] is an approach similar to SeeSoft except that CVSScan is based on a single source code visualization. A program evaluator could easily see the changes performed by a programmer. The focus on one source file provides a finer-grained analysis but it hinders the evaluator's ability to see the programmer's skill across many projects—his worth at a macro level.

Finally, Chronia [6] is a program to track ownership of files based on the amount of modifications made by the developers. The idea is that the developer will become the owner of the file when he contributes more to that file than others. Chronia then creates an ownership map that can be analyzed by managers/evaluators. Since this covers many files, it is a reasonable indicator of the importance/performance of a developer but it lacks the ability to see this from a macro-level. In addition, developers who work on code fixing important parts of it may never get ownership of the file leading the evaluators to have an unfavorable view of them.

2.3 Network Analysis of Software Development

From the descriptions in the previous section one factor should be observed: the current approaches rely on numbers that depend solely on the individual. As new advances in both software development practices and hardware have been made, the size and complexity associated with the development of software has continued to grow. One could argue that any piece of software of interest must be developed by a team of engineers, and it is not uncommon to see team sizes in the hundreds. As software development becomes increasingly more team-oriented so too must the methods used to measure the individual developers. Traditional metrics are quite simple; they observe the individual at the expense of the team. Network measures, on the other hand, take into account the individual's position within the network of developers, and, as a result, can paint a much more accurate picture of each individual's worth within that organization.

There have been studies in the past that have used network measures to examine the software development process. The first, conducted by researchers at Notre Dame University [7], examined the entire Free/Libre/Open Source Software (FLOSS) community as a single network of developers connected by their active participation in common projects. The main goal of this study was to determine if the network of FLOSS developers showed the same characteristics of other communication networks that had been found in previous studies on real networks [8,9]. They did, in fact, find that the FLOSS community does follow a power-law distribution and that the community seems to grow, not as a random network, but rather as a preferentially attached network.

The second study, performed at Syracuse University [10], looked at the internal structure of several projects in the FLOSS community to determine if they followed the widely held belief that FLOSS projects tend to be chaotic and decentralized. The common meme being that FLOSS software is constructed in a "bazaar" style—with little to no real planning and lots of people pitching in where needed—whereas proprietary software is built more like a "cathedral"—with well thought out and carefully architected designs and processes. For each project a network was constructed by linking individual participants to one another if they had participated in any interaction on a common issue through the project's issue tracking software. The study found that the centrality measures across the 120 different FLOSS projects tended to follow a Gaussian distribution. This suggests that the conventional idea that FLOSS projects are highly decentralized and chaotic, like that of the atmosphere of a "bazaar", is not entirely correct. In fact, the FLOSS projects showed a broad scope of social structures ranging from highly centralized to thoroughly decoupled. It was also found that a negative correlation exists between the size of the project and the centrality of that project. Therefore, as the projects grow in size, they also tend to grow in entropy as well, becoming more chaotic and decentralized as membership in them increases. This study seemed to prove that neither depiction of FLOSS software—"cathedral" or "bazaar"—was an accurate representation of all FLOSS projects and that, instead, open-source software projects are wildly varied in how each is developed.

3 General Concepts

3.1 Shortcomings in Traditional Metrics

We have described, in earlier sections, many metrics that are used in industry to evaluate software engineers. Despite these many attempts, the problem of correctly identifying talent (and consequently identifying under-performers) still challenges managers. Such identification processes allow companies to reward developers according to their contribution to the company. Unfortunately, the state-of-the-art in industry is to rely on metrics that are isolated from the context in which they are applied. For instance, the common SLOC does not take into consideration factors such as the difficulty of the code being worked on, the quality of the code generated (i.e. n lines of code are not always equal to n lines of code), or the importance of that source code to the entire project. Due to this lack of metric, managers generally resort to reviews, meetings, and interviews to understand the importance of a developer in the entire context of the company.

In this paper we argue that Network Science [11] can be used to model interactions in a software development company; one can create a social network of developers and use this network to measure the importance of each to the company. It follows that the importance of an individual can, and should, be given by his rank in the network based on approaches such as centrality as well as approaches inspired from his importance to the maintenance of the network stability (as in Random-Targeted attacks in Complex Networks [12,13]).

3.2 Network Measures for Node Rank

Networks are characterized by many measurements which can generally be classified as micro and macro. Macro measurements are characteristics that pertain to the entire network, such as degree distributions, number of communities, and average path length. Contrasting with these we have micro measures which refer to the characteristics of individual nodes and edges in the network and are used to rank these nodes and edges inside the network.

The most basic example of a micro property is the degree of a node, given by $\deg(v)$ and used as a property of nodes since the beginnings of graph theory with Euler. The $\deg(v)$ in a network represents the number of connections a node has to the other nodes where these connections can be directed or undirected. In directed graphs, $\deg(v) = $ $\mathrm{indeg}(v) + \mathrm{outdeg}(v)$, meaning that a node, v, may have both incoming and outgoing edges. Nodes with a high degree are generally thought of as being susceptible effects from information transmitted on the network. In order to normalize the degree values one refers to the degree centrality of a node, which is given by

$$C_d(v) = \frac{\deg(v)}{|V| - 1}. \tag{1}$$

In network analysis it is common to try to find a node that is shallow on the network, meaning that its distance to all other nodes is small. The measure that captures this geodesic distance to all other nodes is defined as the closeness centrality of a node, given by $C_c(v)$. The most accepted definition of closeness centrality is

$$C_c(v) = \frac{1}{|V| - 1} \sum_{s \in V \setminus v} d(v, s), \tag{2}$$

where $d(v,s)$ is the minimum distance between v and s. Note however that it is also common to use an alternative definition for $C_c(v)$ defined as the maximum of all minimum-distances between v and the other nodes.

Betweenness is another node property related to centrality. Informally, it represents how much a node seems to be important in the connection of other nodes. Hence a node with high betweenness appears in many of the shortest paths between pairs of nodes. The definition of betweenness is given by

$$C_b(v) = \frac{1}{\sigma_{st}} \sum_{s,t \in V \setminus v} \sigma_{st}(v), \tag{3}$$

where $\sigma_{st}(v)$ is a count of all the shortest paths that go through node v.

We argue in this paper that centrality measures are better at evaluating developers than current approaches based on the individual. These measures provide managers with a sense of how the network of developers depends on a particular individual. Even more interesting, we show that we can use a network of files (see next Section) to evaluate the importance of a developer. In order to confirm our hypothesis, we perform many experiments simulating developers being removed from the project (deleted from the network) according to the centrality measures done here as well as the individual approaches described earlier. Our results demonstrate that the removal of developers based on individual metrics such as SLOC count is either equivalent to or worse than removing developers via network centrality measures.

4   Case Study: The Python Open-Source Project

The Python programming language was chosen as the focus of this study for several reasons. The first of which was the authors' familiarity with the language and its surrounding community. This familiarity brought with it two main advantages: a slight understanding of the code base, and a knowledge of who are the "linchpin" developers in the community. The latter advantage acts as some level of qualitative measurement of the success of the methods used to determine developer importance.

Next, we chose Python for its size and structure. According to the study done by Crowston and Howison [10], Python, relative to the other 119 FLOSS communities in the study, has a large and decentralized developer community. We chose Python specifically because we wanted a community that would rival, in size, many of the development teams that one would see at some of the larger successful software companies such as Google, Microsoft, and Apple. In addition to the size, the decentralized nature of the project seemed to fit the architecture that one would see in many companies where the software is developed by large numbers of small groups. It is due to these features that we felt that the Python project would stand in as a rough analog for a typical proprietary project in a large company.

We constructed three separate networks in order to study the dynamics of the Python software development community. In the next few sections we will look at each network in turn and discuss the benefits that each provides.

4.1 Source-Code Dependency Network

The Source-Code Dependency Network is essentially a graph-based view of the software's architecture. In other words, we have assembled a directed graph based on the file dependencies (i.e., #includes since Python is written in C) inside of each source code file. There were a couple of interesting properties to take note of in this network. First, the scaling factor of the power law given by $\alpha \approx 3.5 \pm 0.1$, when the network is analyzed as undirected. This suggests that the degree distribution for this network does indeed follow a power-law distribution. This is a fact that we can easily see in Figure 1. The major hubs in the graph are highlighted in red (and are larger in size) and labeled with the file's name. As you can see there are 5 main hubs (a few smaller ones, but only those with a degree of 10 or higher are labeled here).

Next, the clustering coefficient is zero—there is no clustering at all.This is somewhat expected as a clustering coefficient above 0 would suggest that there are circular references in the software, a property that is usually ardently avoided in most software development practices. So, a zero clustering coefficient can be a sign of a well-architected system.

The main argument in this paper is that metrics like SLOC or bug-counts are not very useful in evaluating a developer's value to the project. Hence, we use the Source-Code Dependency Network to show that important files (like the hubs in Figure 1) are not strongly correlated to the number of developers working on them or the issues (bugs) found on them. The data supports this view as the Pearson correlation for each was found to be 0.03 ± 0.05 and 0.01 ± 0.02 respectively, in other words, no correlation was found to exist. Hence we extracted the bugs that were reported as fixed in the SCM (Source Configuration Management) logs and overlaid that information on our network. We did the same for the developer data. This information can be seen in Figure 2. The two networks show the bug data, Figure 2(a), and the developer data, Figure 2(b). The data shown is the number of bugs and developers per file as a heat-map where blue (cold) nodes represent files with relatively few bugs and developers and red (hot) nodes representing files with many bugs and developers; to make this even clearer, the size of the nodes are also proportional to the number of bugs and developers associated with each file. As can be seen from the networks, the number of bugs per file tends to correspond to the number of developers per file. In other words, the more developers that worked on a file, the better chance that more bugs would be reported on that file as

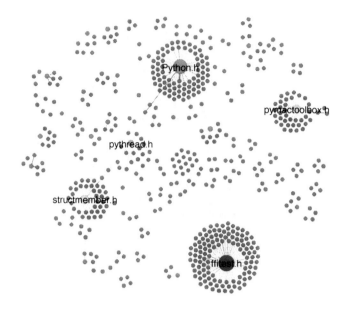

Fig. 1. Source Code Dependency Network

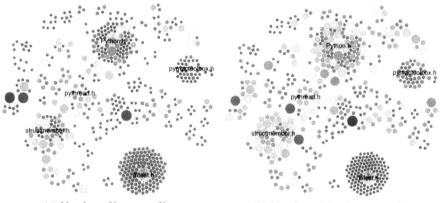

(a) Number of bugs per file   (b) Number of developers per file

Fig. 2. Bug and Developer Data. A heat-map for the color of the nodes as well as their size represent the numbers of bugs/file and developers/file.

well. The data supports what we can intuitively tell from the network, as the Pearson correlation coefficient for the data was $\approx 0.70 \pm 0.05$ representing a high correlation between the two. It is also worth mentioning that the distribution of the number of bugs found per file and the number of developers per file follows a power law. Our findings suggest that both distributions (bugs/file and developers/file) have scaling given by 1.95 ± 0.02 and 2.0 ± 0.05 respectively.

Recall however that the main claim in this paper relates to using network to evaluate developers. The results in the previous section indicates that bug counts and developer counts do not strongly correlate to the importance of a node, given in Figure 1 by hub nodes. To study the importance of a developer we have created another network from the SCM logs. We generated a bipartite network with source code files and developers as nodes, where the edges represent the fact that a developer has worked on a file. This bipartite graph allow us to explore either a file projection or a developer projection of the bipartite network.

4.2   Developer Network

The developer network is an undirected projection of the bipartite network in which nodes represent developers in the community and edges represent collaborations; an edge was drawn between two nodes if both developers had worked on at least one common file. A weight was then assigned to each of these edges based on the number of common files on which the two developers had worked, thus, strong and weak collaborative relationships could be easily distinguished.

The developer network is, in essence, an indirect social network—indirect because developers are not required to know each other in the real world to essentially be collaborators in the software development project. However the network topological characteristics are not true for social networks since they are just projections of the real (bipartite) network. For instance, the developer network does not have a strong power-law degree distribution. This might also happen due to the unique fact that Python is an open-source project where contributors (165 in our case) are inclined to look at most files and hence be connected to most other developers in this projection. Some of these links however are sporadic. If we consider the distribution of weights as part of node properties (as argued by Opsahl et al. [14]) the network behaves as do many other real networks.

In order to evaluate the importance of a developer, we have looked at the size of the largest connected component in the network (aka Giant Component). The premise here is that if a user is important, she should be missed in the network when removed. This "missing" factor occurs if the removal of the developer from the network causes the network to become more disconnected (i.e., the size of the Giant Component decreases significantly).

Figure 3 is used to demonstrate that metrics based on the individual are not very realistic in identifying relevant developers. To demonstrate this fact we have removed developers from the network randomly and by the number of commits the developer had to the project (a commit occurs when a developer has produced a new version of the source code; a good approximation for SLOC). Figure 3 shows that the random removal and commit-count removal (highest commit-count first) causes a similar effect to the giant component—it decreases in size somewhat linearly. This result demonstrates that commit-count is as good (or as bad) as a metric to identify talent in a software development process as a random identification. Figure 3 also shows our proposed approach in which talented developers are removed according to centrality measures in the network. Note that they all provide a significantly better measure of talent since the network degenerates faster after about 30% of the nodes are removed. This means that once the developer network is formed one can decide on talent based on the centrality, closeness or degree of the developer.

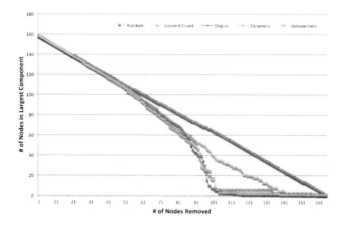

Fig. 3. Behavior of the giant component in the developer network as nodes are removed from the network

5   Conclusion and Future Work

In this paper we propose that current methods for evaluating software developers within an organization are poor indicators of the overall importance of the developer. Metrics that have been in use since the creation of the software engineering discipline are still largely in use today despite their well known inadequacies. As a solution, we propose the use of network measures of centrality as a better indicator of how vital each developer is to the success of their organization. In this study, we have chosen a large, well-known, open source project with a very active developer community with which to test our hypothesis. The previous section discusses the results of this study in detail, but in short, we found that network measures were indeed a much more accurate measure of the importance of an individual developer within a development community.

There are two main areas on which we would like to concentrate our efforts for future work. The first is simply solidifying the results we have attained thus far by experimenting with several more projects in the Open Source community to make sure that our results are consistent across a large number of different projects and not specific to only the Python community. Testing these measurements on a proprietary project, we feel, would also help to cement the results found in this study and is an option that we are actively pursuing at this time. The second area we would like to focus on is the dynamic nature of network measurements and how important it may be in determining developer importance. To illustrate this, consider a community of several teams of developers where each team works on a specific area of the software with some overlap between teams. As members of the community come and go, for whatever reason, the importance of individuals within the organization changes. For example, if several developers on a single team leave at once, the importance of the remaining members of that team should grow since the loss of these individuals would open up a structural hole in the community. Since network centrality measures take into account the structure of the network, they tend to change as the structure of the network changes, whereas measures such as code and commit count remain constant no matter how much the surrounding environment changes. In this study we focused on measuring an individual's importance with respect to the network as it currently exists. For future work we would like to examine how changes in the community structure effect the importance of individuals within it and how well our method captures those changes as opposed to the static methods used in current practice.

References

1. Naur, P., Randell, B.: Software engineering: Report of a conference sponsored by the nato science committee. Technical report, North Atlantic Treaty Organization (1968)
2. Fenton, N.E., Neil, M.: Software metrics: successes, failures and new directions. Journal of Systems and Software 47(2-3), 149–157 (1999)
3. Sheetz, S.D., Henderson, D., Wallace, L.: Understanding developer and manager perceptions of function points and source lines of code. Journal of Systems and Software 82(9), 1540–1549 (2009)
4. Eick, S.C., Steffen, J.L., Sumner, E.E.: Seesoft-a tool for visualizing line oriented software statistics. IEEE Transactions on Software Engineering 18(11), 957–968 (1992)
5. Voinea, L., Telea, A., Wijk, J.: CVSscan: visualization of code evolution. In: SoftVis 2005: Proceedings of the 2005 ACM Symposium on Software Visualization (May 2005)
6. Gîrba, T., Kuhn, A., Seeberger, M., Ducasse, S.: How developers drive software evolution. In: Eighth International Workshop on Principles of Software Evolution, pp. 113– 122 (2005)
7. Madey, G., Freeh, V., Howison, J.: The open source software development phenomenon: An analysis based on social network theory. In: Proceedings of AMCIS 2002 (2002)
8. Watts, D.J., Strogatz, S.H.: Collective dynamics of small world networks. Nature 393, 440–442 (1998)
9. Barabási, A.-L., Albert, R.: Emergence of scaling in random networks. Science 286(5439), 509–512 (1999)
10. Crowston, K., Howison, J.: The social structure of free and open source software development. First Monday 10(2) (2005)
11. Lewis, T.G. (ed.): Network Science: Theory and Applications. Wiley, Chichester (2009)
12. Pastor-Satorras, R., Vespignani, A.: Immunization of complex networks. Phys. Rev. E 65(3), 036104 (2002)
13. Barabasi, A.-L., Bonabeau, E.: Scale-free networks. Scientific American, 50–59 (2003)
14. Opsahl, T., Agneessens, F., Skvoretz, J.: Node centrality in weighted networks: Generalizing degree and shortest paths. Social Networks 32(3), 245–251 (2010)

Hybrid Complex Network Topologies Are Preferred for Component-Subscription in Large-Scale Data-Centres

Ilango Sriram and Dave Cliff

Department of Computer Science, University of Bristol,
Bristol, UK
{ilango,dc}@cs.bris.ac.uk

Abstract. We report on experiments exploring the interplay between the topology of the complex network of dependent components in a large-scale data-centre, and the robustness and scaling properties of that data-centre. In a previous paper [1] we used the *SPECI* large-scale data-centre simulator [2] to compare the robustness and scaling characteristics of data-centres whose dependent components are connected via Strogatz-Watts small-world (SW) networks [3], versus those organized as Barabasi-Albert scale-free (SF) networks [4], and found significant differences. In this paper, we present results from using the Klemm-Eguiliz (KE) construction method [5] to generate complex network topologies for data-centre component dependencies. The KE model has a control parameter $\mu \in [0,1] \in \mathbf{R}$ that determines whether the networks generated are SW ($0 < \mu << 1$) or SF ($\mu = 1$) or a "hybrid" network topology part-way between SW and SF ($0 < \mu < 1$). We find that the best scores for system-level performance metrics of the simulated data-centres are given by "hybrid" values of μ significantly different from pure-SW or pure-SF.

1 Introduction

Modern ultra-large-scale data-centres appear, from the outside at least, to be highly regimented and regularly-structured engineering artefacts. Aisle after aisle of racks, each rack being a vertical frame housing a number of chasses, each chassis housing a regular arrangement of thin computer mother-board units: the "blade-servers" that make up the data-centre's computing infrastructure. In this paper, we show that *hybrid* complex network topologies of dependencies between the computing components that make up a data-centre service may bring benefits. We emphasise the hybrid nature of the complex network topologies here because the results published in this paper demonstrate that "non-standard" network constructions are best: these topologies are non-standard in the sense that they are neither purely scale-free (SF) nor purely small-world (SW), but rather part-way between the two.

Data centres are becoming ever-larger as the operators seek to maximise the benefits from economies of scale. With these increases in size comes a growth in system complexity, which is usually problematic. The growth in complexity manifests itself in two primary ways. The first is that many conventional management techniques that work well when controlling a relatively small number of data-centre nodes scale much worse than linearly and hence become impracticable when the number of nodes

L. da F. Costa et al. (Eds.): CompleNet 2010, CCIS 116, pp. 130–137, 2011.

under control increases by two or three orders of magnitude. The second is that the very large number of individual independent hardware components in modern data centres means that, even with very reliable components, at any one time it is reasonable to expect there always to be one or more significant component failures (so-called "normal failure"). Despite this, guaranteed levels of performance and dependability must be maintained. For an extended discussion of the issues that arise in the design of warehouse-scale data-centres, see [6].

Predictive computer simulations are used in almost all of current engineering practice, to evaluate possible designs before they go into production. Simulation studies allow for the rapid exploration and evaluation of design alternatives, and can help to avoid costly mistakes. In microelectronics e.g., the well-known *SPICE* circuit-simulation system [7] has long allowed large-scale, highly complex designs to be evaluated, verified, and validated in simulation before the expensive final stage of physical fabrication. Despite this well-established tradition of computational modelling and simulation tools being used in other engineering domains, there are currently no comparable tools for simulating cloud-scale computing data-centres. The lack of such tools prevents the application of rigorous formal methods for testing and verifying designs before they go into production. Put bluntly, at the leading edge of data-centre design and implementation, current practice is much more art than science, and this imprecision can lead to costly errors.

As a first step in meeting this need, we have developed *SPECI* (Simulation Program for Elastic Cloud Infrastructures). Clearly, it would require many person-years of effort to bring *SPECI* up to the comprehensive level of *SPICE* or of commercial CFD tools. We are currently exploring the possibility of open-sourcing *SPECI* in the hope that a community of contributors then helps refine and extend it.

The first paper discussing *SPECI* [2] gave details of its rationale and design architecture that will not be repeated here. In that first paper, results were presented from simulation experiments that had been suggested by our industrial sponsor, Hewlett-Packard Laboratories. The specific area of inquiry is large-scale data-centre middleware component-status subscription-update policies. The status of data-centre components may change as they fail, or as policies are updated. Within the data-centre, there will be components that work together and need to know the status of other components via "subscriptions" to status-updates from those components. In [2] we used a first-approximation assumption that such subscriptions are distributed randomly across the data centre. That is, the connectivity of the network of subscription dependencies within the data-centre was, formally, a random graph. In [1], we explored the effects of introducing more realistic constraints to the structure of the internal network of subscriptions. We explored the effects of making the data-centre's subscription network have a regular lattice-like structure, and also the effects when the networks had complex semi-random structures resulting from parameterised network generation functions that create small-world [3] and scale-free [4] networks. We showed that for distributed middleware topologies, varying the structure and distribution of tasks carried out in the data centre can significantly influence the performance overhead imposed by the middleware.

In this paper we inspect the use of hybrid complex network topologies as basis for the structure and distribution of tasks carried out in the data centre. We use the Klemm-Eguiliz (KE) construction method [5], which has a parameter $\mu \in [0,1] \in \mathbf{R}$ that determines whether the networks generated are SW ($0<\mu\ll1$) or SF ($\mu=1$) or a "hybrid" network topology part-way between SW and SF ($0<\mu<1$). We find that the best scores for system-level performance metrics of the simulated data-centres are given by "hybrid" values of μ significantly different from pure-SW or pure-SF.

The structure of this paper is as follows. In Section II we give further details of *SPECI*, sufficient for the reader to comprehend the new results presented in this paper. In Section III we validate our modifications to KE that turned it into a directed-graph generator. In Section IV, we summarise some of the results presented in [1] against which we then compare the outputs from the directed-KE hybrid complex subscription network. Our conclusions are given in Section V.

2   Explanation of SPECI

A software layer (so-called "middleware") that is responsible for job scheduling, load-balancing, security, virtual network provisioning, and resilience binds the components of the DCs together, and is the DC's management layer. Scalability requires the performance not to rapidly degrade as the number of components increases, so that it remains feasible to operate in the desired size range [8]. Yet it is unlikely that all properties in middleware will scale linearly when scaling up the size of DCs.

Because the middleware's settings and available resources change very frequently, it needs to continuously communicate new policies to the nodes. Traditionally middleware manages its constituent nodes using central control nodes, but hierarchical designs scale poorly. Distributed systems management suggests controlling large DCs using policies that can be broken into components for distribution via peer-to-peer (P2P) communication channels, and executed locally at each node. P2P solutions scale better, but can suffer from problems of timeliness (how quickly updated policies will be available at every node) and of consistency (whether the same policies are available and in place everywhere).

The simulation contains a number (n) of nodes or services connected through a network. Each of these nodes can be functioning (alive) or not, and state changes occur at a change-rate f. To discover the aliveness of other nodes, each node provides an arbitrary state to which other nodes can listen. When the state can be retrieved the node is alive, otherwise it is not. Every node is interested in the aliveness of some of the other nodes, the value of "some" being a variable across all nodes. Each node maintains a subscription list of nodes in whose aliveness it is interested. *SPECI* provides a monitoring probe of the current number of inconsistencies, and the number of network packets dealt with by every node, per unit time (here every second). For now, the simulator assumes uniform costs for connecting to other nodes, but we intend to explore varying the connection costs in a meaningful way, in future work.

Initially, we observed the number of nodes that have an inconsistent view of the system. A node has an inconsistent view if any of the subscriptions that node has contains incorrect aliveness information. We measure this as the number of inconsistent nodes, observed here once per Δt (=1sec). After an individual failure or change occurs, there are as many inconsistencies as there are nodes subscribed to the failed node. Some of these will regain a consistent view within Δt, i.e. before the following observation, and the remaining ones will be counted as inconsistent at this observation point. If the recovery is quicker than the time to the next failure, at the subsequent observations fewer nodes will be inconsistent, until the number drops to zero. When the update retrieval method requires aliveness data to be passed on, more hops would make us expect more inconsistencies, as out-dated data could be passed on. This probing was carried out while running SPECI with increasing failure rates and scale, and using each of these combinations with each of our four protocols. We scaled n though DC sizes of 10^2, 10^3, 10^4 and 10^5 nodes. We assume that the number of subscriptions grows slower than the number of nodes in a DC, and so we set the number of subscriptions to $n^{0.5}$ per node.

For each of these sizes a failure or change rate distribution f was chosen such that on average in every minute 0.01%, 0.1%, 1%, and 10% of the nodes would fail. A gamma distribution was used with coefficients that would result in the desired rate of failures. Each pair of configurations was tested over 10 independent runs, each lasting 3600 simulation time seconds, and the average number of inconsistencies along with its standard deviation, maximum, and minimum number were observed. The half-width of the 95% confidence intervals (95% CI) was then calculated using the Student's t-distribution for small or incomplete data sets.

Formally, the transitive P2P protocol is represented by a directed acyclic graph (DAG) with these distribution topologies. State-changes are introduced to this DAG at a rate f, and load and inconsistencies during the propagation of the state changes are measured. The work reported here continues investigating the trade-off between load and inconsistencies between SW and SF, and continues exploring further subscription topologies by investigating hybrid complex network topologies, and show benefits such hybrid topologies can bring as distribution patterns for services in large-scale data centres. The following subsection explains the complex network topologies used.

2.3 Complex Network Topologies

The original algorithms for constructing these styles of network are for undirected graphs; but the *SPECI* subscription network requires a directed graph. Here we describe the directed implementations used as well as the applicability of the topology.

Small world (SW) undirected networks were first constructed by Watts and Strogatz using an algorithm described as rewiring a ring lattice [3]. Kleinberg has proposed a directed version of SW networks, which starts from a two-dimensional grid rather than a ring in his work searching for a decentralised algorithm to find shortest paths in directed SW [9]. However, here we stay closer to the original algorithm by Watts and Strogatz with the only modification that the initial condition is a ring lattice with outgoing directed edges to all k nearest neighbours, k being the number of subscriptions in our model. Thus, between neighbouring edges there will be two edges, one in each direction. As rewiring probability $p=0.1$ was used, a value Watts and Strogatz showed to be large enough to exhibit short path lengths, and at the same time small enough to exhibit a high transitivity of the network [3]. SW networks have a high clustering coefficient or transitivity in the directed case (i.e. many of the neighbours are themselves neighbours) similar to regular lattice graphs, but at the same time they have low diameters (i.e. short average path lengths) as found in random graphs. Watts and Strogatz [3] showed that networks with such properties arise naturally in many fields and are commonly found in natural phenomena. In a data centre, we could imagine the subscription graph having a SW distribution when components are initially placed close to each other, and over time change their location or their functionality (e.g. as a result of load-balancing) thereby turning from local into long range contacts.

Scale-free (SF) undirected networks were first constructed by Barabási and Albert (BA) by growing a network with preferential attachment [4, 10]. Unfortunately, the Barabási-Albert model cannot be directly transferred into a directed network. Several attempts exist to model a directed version. They mostly differ in how the initial network that is to be grown is generated, and in how the problem is overcome that directed SF networks are often acyclic, unlike real world networks. For a discussion of weaknesses of directed versions of the Barabási Albert model see the Newman's review of complex networks [11]. For our work, the implementation proposed by Yuan and Wang [12] is used. This starts with a fully connected network, and then grows it using preferential attachment. This results in a citation-network with feed-forward characteristics, where each new node is connected pointing towards the existing nodes. However, with a small probability p the direction of edges is inverted, to allow the generation of cycles. We used $p=0.15$, which was a value under which Yuan and Wang found the network to be "at the edge of chaos" when transformed into Kauffman's NK model [13]. SF networks have low diameters, and been found in many complex networks in the real world, too. For instance, many aspects of the internet are SF networks. SF networks could have applicability for subscriptions when there are components with different importance for other nodes, for instance.

Hybrid complex networks have all three characteristics of the previous complex networks, namely low diameters, a high clustering coefficient, and a power-law distribution, from which the SW and SF each exhibit only two but lack one [5]. There are two algorithms for building hybrid complex networks: The one by Klemm and Eguiluz (KE) [5] is used in this work, and there is another implementation by Holme and Kim [14] that is not studied here. The KE model starts with a fully connected network of k nodes, which are all marked as "active", and is grown by iteratively adding nodes that connect to all active nodes and then take the active token from one of the previously active nodes, where the probability of deactivation is proportional to the invers of its current degree. However, with probability $\mu \in [0,1] \in R$ each of the introduced edges becomes a long-range connection that is connected using plain preferential attachment as in the BA model. Thus, for $\mu=1$ the model is identical to BA, for $\mu=0$ it is a network with high clustering and power-law degree distribution but without low diameter, and for "hybrid" values part-way between $(0<\mu<1)$, all three characteristics are present and the cross-over between the models can be studied.

The original KE implementation is an undirected model, as was the case with BA. We applied the same method Yuan and Wang [12] used for plain BA to create directed networks. A directed feed-forward graph of the KE model is grown, and an "invert probability" determined the likelihood that some of the nodes would be inverted to allow the creation of cycles in the graph. Section 3 shows that the characteristics of the KE are still valid after this modification, with transitivity replacing clustering coefficient observations.

3 Growing Directed Scale-Free Small-Worlds

In Subsection 2.4 the KE algorithm to generate hybrid complex networks was described. In this section we show that the directed version we use has the same properties as the original KE model, and look at the behaviour when modifying the invert probability. Fig. 1 shows the effect of introducing $\mu \ll 1$ of random links to the highly clustered model. The average path length drops rapidly with the introduction of these links, while the transitivity is preserved until μ reaches the order of 1. The graph in Fig. 1 has the same shape as the graph in the original undirected version of KE, i.e. Fig. 1 in [5]. Note that values for transitivity replace the values for clustering coefficient, as clustering is not defined for directed networks. Fig. 2 shows the effect of varying the invert probability for generating the directed version of the graph. Transitivity and Path length appear to change at similar rates throughout the values for the invert parameter.

4 Results

The results from [1 and 2] showed that the distribution topology of collaborative components in a data-centre has an impact on the performance of the middleware, and needs to be taken into account when tuning and selecting the topology of the middleware-dependency network. This motivates us to search for preferable distribution topologies. It also demonstrates the value of rigorous simulation tools that allow exploratory studies of planned designs for large-scale data centres, identifying effects that were not anticipated or deliberately accounted for at design phase. We found that the transitivity in the small-world subscriptions caused benefits in the load generated, but also caused the existence of many more inconsistent nodes, as out of date information was passed on. The use of the KE model throws further light on the influence of path lengths and transitivity to the middleware performance. Fig. 3 shows the outcome of *SPECI* runs where the subscriptions are distributed according to a KE network: The level of inconsistencies in the graphs with larger n and f drops with increasing μ up to a certain key value μ^*; after this threshold value the inconsistencies rise sharply. Note that the value for $\mu=1$ differs from the value in the previous section, because the algorithm here starts with an initial set of k connected nodes, while the previously used implementation of BA started with an initial set of $(k-1)$ unconnected nodes which were then connected to and from the k-th element.

This outcome leads to two assumptions: First, the best scores for system level performance occur under hybrid topologies that are neither entirely SW nor entirely SF. Second, the inconsistencies fall as the path lengths drop. When μ grows to the extent that the transitivity drops, a sharp rise in inconsistencies can be observed. This contradicts our previous belief that the transitivity caused the higher levels of inconsistencies in our previous results from plain SW. Based on these insights we can recommend the subscription topology in a data centre to have both high transitivity and low path lengths.

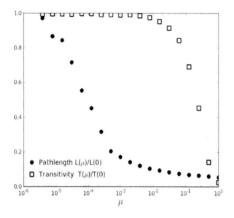

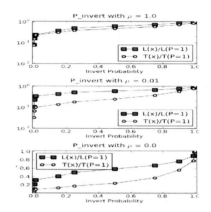

Fig. 1. Small-world effect in directed scale-free networks. Tuning the factor μ in the directed version of KE shows the same effects to Path length and Transitivity as the undirected version [5] does to path length and clustering coefficient. n=10^4 nodes with average outdegree k=20 where used, as in Fig. 1 of the original KE publication [5], and the invert probability was set to 0.15.

Fig. 2. Unlike the factor μ, the invert probability used to create the directed model does not exhibit comparable effects on Path length and Transitivity. In BA (μ=1) Yuang and Wang reported an invert Probability around 0.15 to reflect common scale-free networks.

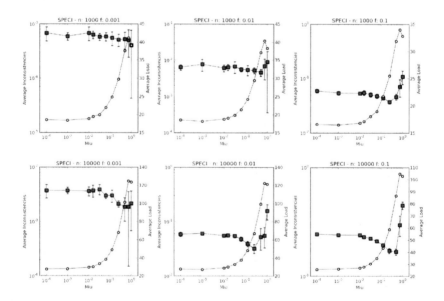

Fig. 3. Increasing μ from small-worlds (μ=0) to scale-free networks (μ=1) for DC sizes of 10^3 and 10^4 with change rates of 0.1%, 1% and 10%. The load increases for larger μ. The inconsistencies, however, drop until μ=0.2, and rise sharply for larger μ. This shows that rather than a pure BA distribution of subscriptions in a data centre, a hybrid model would be desirable.

5   Conclusion

In this paper we have demonstrated the use of rigorous simulation tools to allow exploration of planned designs for large-scale data centres. Such simulations can be used to identify (and hence avoid) unanticipated and undesirable behaviours. Our results presented here indicate that *hybrid* complex network topologies of dependencies between the computing components may bring benefits. The best scores for global performance levels were found for values that were neither purely SW nor BA SF networks. Finally, the KE model has helped to identify the level of contribution of path length and transitivity to the results in our previous results.

References

1. Sriram, I., Cliff, D.: Effects of Component-Subscription Network Topology on Large-Scale Data Centre Performance Scaling. In: Calinescu, R., Paige, R., Kwiatkowska, M. (eds.) Proceedings of the 15th IEEE International Conference on Engineering of Complex Computer Systems (ICECCS 2010), pp. 72–81. IEEE Computer Society Press, Los Alamitos (2010)
2. Sriram, I.: SPECI, a Simulation Tool exploring cloud-scale Data Centres. In: Jaatun, M.G., Zhao, G., Rong, C. (eds.) Cloud Computing 2009. LNCS, vol. 5931, pp. 381–392. Springer, Heidelberg (2009)
3. Watts, D., Strogatz, S.: Collective Dynamics of 'Small-World' Networks. Nature 393, 440–442 (1998)
4. Barabási, A.-L., Albert, R.: Emergence of Scaling in Random Network. Science 286, 509–512 (1999)
5. Klemm, K., Eguiluz, V.M.: Growing Scale-Free Networks with Small-World Behavior. Phys. Rev. E 65 (2002)
6. Barroso L., Hölzle, U.: The Datacenter as a Computer: An Introduction to the Design of Warehouse-Scale Machines Synthesis Lectures on Computer Architecture 4 (2009)
7. Nagel, L.W.: SPICE2: A Computer Program to Simulate Semiconductor Circuits. Technical Report No. ERL-M520, University of California, Berkeley (1975)
8. Jogalekar, P., Woodside, M.: Evaluating the Scalability of Distributed Systems. IEEE Transactions on Parallel and Distributed Systems 11(6), 589–603 (2000)
9. Kleinberg, J.: The Small-World Phenomenon: An Algorithmic Perspective. In: 32nd ACM Symposium on Theory of Computing (2000)
10. Albert, R., Barabási, A.-L.: Statistical Mechanics of Complex Networks. Rev. Mod. Phys. 74(47) (2002)
11. Newman, M.E.J.: The Structure and Function of Complex Networks. SIAM Review 45, 167–256 (2003)
12. Yuan, B., Wang, B.: Growing Directed Networks: Organization and Dynamics. New J. Physics 9(8), 282–290 (2007), http://arxiv.org/pdf/cond-mat/0408391
13. Kauffman, S.: The Origins of Order. OUP, Oxford (1993)
14. Holme, P., Jun Kim, B.: Growing Scale-Free Network with Tunable Clustering. Phys. Rev. E 65, 026107 (2002)

A Network-Centric Epidemic Approach for Automated Image Label Annotation

Mehdy Bohlool, Ronaldo Menezes, and Eraldo Ribeiro

Florida Institute of Technology, 150 W University Blvd., Melbourne, FL 32901, USA
mbohlool2009@my.fit.edu, {rmenezes,eribeiro}@cs.fit.edu

Abstract. Automatically organizing and searching images by their content in large image datasets are major goals of Web-based search engines. Currently, these goals are accomplished by associating metadata information to each image in the database. In this paper, we investigate the use of network sciences to implement a metadata-propagation mechanism for images. We begin by creating a network of images connected by a part-based appearance-similarity measure, and propose an epidemiology-inspired model for metadata propagation. Our experiments show that organizing images as a network helps us label a large number images in the dataset in an economical way, i.e., with few manual metadata annotations.

1 Introduction

The amount of non-textual information available on the Web is increasing significantly. While the Web is highly populated by images, music, and video data, search engines are still dependent on the availability of manually-added textual metadata associated with these non-textual documents. However, manually including metadata to every non-textual Web document is clearly unfeasible. It is evident that the capability for searching non-textual content is likely to become a crucial part of modern Web search engines.

In this paper, we approach the problem of automatically propagating metadata labels in partially labeled image datasets. We consider that image databases can be represented by a network in which images are connected to one another by a part-based appearance-similarity measure. Based on this network configuration, we propose a metadata-propagation mechanism inspired by an epidemiology model.

Real-world networks can be broadly classified into two main non-disjoint classes named Scale-Free Networks [2] and Small-World Networks [15]. Informally, scale-free networks are characterized by having the majority of nodes with small degree and very few nodes with very large degree (i.e., a power-law degree distribution). Small-world networks, on the other hand, present inter-node connections of short path lengths while the network itself contains many highly connected clusters of nodes (i.e., high clustering coefficient).

Closely related to our paper is the study of networks for the understanding of how viruses spread in the real world [11]. More specifically, the development of effective immunization models that target a small subset of nodes in a network [12]. What is important for us here is to understand that certain nodes in the graph have a special role that may be used in the process of stopping or starting an epidemic.

L. da F. Costa et al. (Eds.): CompleNet 2010, CCIS 116, pp. 138–145, 2011.

Organizing images as networks is not novel. Deng Cai et al. [4] constructed an image network using images from websites by using a combination of visual appearance, textual information, and link information. Their method grouped images into semantic classes to improve users' Web searching experience.

The key to organizing images as networks is to find a similarity measure to relate the images. The scale-invariant feature transform (SIFT) [9] is an effective technique for finding similar features in images. Inter-image similarity measures based on SIFT have been proposed for image categorization [7,13]. Once the image network is at hand, we want to use the network to propagate annotation. Von Ahn et al. [14] were one of the first to look at annotation; they designed a game to make it less burdensome (and more fun) for people to annotate images manually. Hare et al. [8] proposed an auto-annotation system based on the propagation of keywords which is the main inspiration for our propagation proposal.

2   Building the Network of Images

We represent an image dataset as a network defined by an undirected weighted graph $G = (V, E)$ where the vertices $V = \{v_1, \ldots, v_n\}$ are the images, and an edge $e_{ij} \in E$ indicates that images v_i and v_j are connected. The edge strength is the affinity weight between the two images given by a function w that associates a numerical label $w(e_{ij})$ with that edge. In our method, w is defined in terms of the number of consistent matches obtained by comparing local image regions extracted from the two images. Here, the regions are automatically extracted by SIFT [9]. This feature detector is able to describe small image regions or patches (i.e., local appearance) in terms of their intrinsic scale and dominant orientations of pixel intensity gradients (Figure 1). Once normalized with respect to both scale and orientation, SIFT descriptors can be compared in a scale and rotation-invariant manner.

To address the problem of spurious matches, at each detected image location, we combine neighboring matching pairs to form a local set of features for each location. Let $m_k = \{(x_1^k, y_1^k), (x_2^k, y_2^k)\}$ and $m_l = \{(x_1^l, y_1^l), (x_2^l, y_2^l)\}$ represent two pairs of matching points between images k and l, respectively. We say that m_k and m_l are *neighboring matches* if the Euclidean distance between their endpoints on each image is less than a pre-defined threshold τ_N (e.g., $\tau_N = 10$ pixels). We perform this neighborhood check for all matches produced by the SIFT detector, keeping only those matches that have four or more neighboring matches. If n_t is the total number of neighboring SIFT matches between two images, we define the appearance similarity between images v_i and v_j as:

$$w(e_{ij}) = 1 - \exp\left(-\frac{n_t^2}{\sigma^2}\right), \tag{1}$$

where we selected $\sigma = 25$ to provide an approximate fit to the distribution of inter-image similarity measures ($w \in [0, 1]$). Finally, the network is built by performing a pairwise comparison for all images, and assigning edge weights according to Equation 1. In our experiments, we avoided the creation of fully connected networks by removing links with strength smaller than 4.

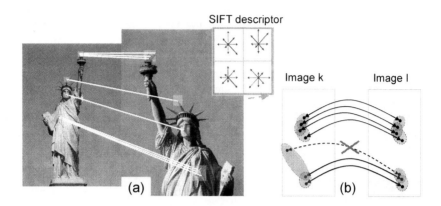

Fig. 1. Image similarity using SIFT matches. (a) SIFT matches and zoomed view of SIFT descriptor as a histogram of local gradient orientations. (b) SIFT pairs that exceed neighborhood size threshold are removed. Image similarity is a function of the total of remaining neighboring pairs.

3   Epidemic-Based Annotation Model

Our goal is to maximize the propagation of image labels through the network of connected images. Our label-propagation method uses a mechanism analogous to real-world virus epidemics and is hence influenced by two main issues: *(i)* how the spread takes place (label propagation), and *(ii)* where the infection starts (choice of an initial target).

Label Propagation. We define β_{ij} as the spreading rate, or probability, that a label will propagate from node i to node j. This rate is given by:

$$\beta_{ij} = \begin{cases} c, & \text{if using uniform propagation} \\ w(e_{ij}), & \text{if using non-uniform propagation.} \end{cases} \tag{2}$$

We use both uniform and non-uniform propagation approaches. The uniform propagation uses a constant infection rate, a process analogous to real-world viruses. In our experiments, we set $\beta_{ij} = 0.5$, and current image labels are iteratively propagated through the network until all connected nodes are reached.

For the non-uniform propagation case, we define a spreading rate for each link as the similarity between image i and image j, i.e., node i propagates a label to a neighboring node j with probability $\beta_{ij} = w(e_{ij})$. Figure 2 shows a simple label-propagation example.

Choosing an Initial Node. As we argued before, image annotation is an expensive process whether it is manual or automated. Propagation efficiency is also influenced by the location where the propagation starts. The first and obvious way to chose a node to start the propagation is the use of a random process. Nodes in the network are chosen at random and labeled manually. Although this is unlikely to produce the best results it is

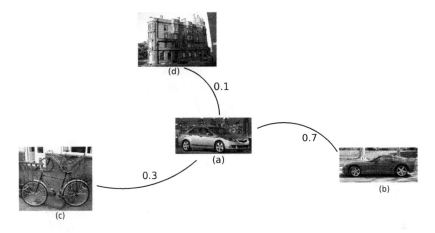

Fig. 2. Example of non-uniform propagation. Node (a) is manually annotated with the label "car". Neighboring nodes have different probabilities of receiving that label based on the normalized SIFT values for the similarity between (a) and the other images.

important to allow us to better evaluate other methods. We focus however on identifying nodes in the network that are considered important. To define importance one generally looks at the number of links a node has. If an image is highly connected to others it is reasonable to assume it can propagate its labels to more images. We refer to this approach as highest-degree targeting (HDT).

It is known however that high-degree nodes are not the only important nodes in a network [6]. There are alternatives that look beyond the number of connections. PageRank [3] is a concept used by Google to rank web pages based on their importance on the Web. Highly ranked pages tend to have neighbors that are also highly ranked. Thus, PageRank describes the probability of reaching a node (webpage) by a random crawler of the network. It depends not only on the degree, but also of how strong the links are, as well as the rank of the node's neighbors. The more popular a node is in the network, the more likely a random user will find it. We use PageRank to identify nodes in the image network that are candidates to start our image propagation. We call this approach highest-ranked annotation (HRT).

4   Experimental Results

The goal of our experiments is to show how our label-propagation model allows for improved annotation capabilities of unlabeled or partially labeled image datasets. First, we created an image network by sampling from a large unlabeled image database using the approach of Section 2. Then, we performed a topological characterization of our network to demonstrate that we are dealing with a complex network. Last, we compared different targeting annotation methods on the image network using our label-propagation models.

Network Creation. We began by creating an image network by randomly selecting images from the Microsoft Cambridge Image Dataset[1]. We selected a sample of 516 images. Once the images were at hand, we ran SIFT to detect and describe points of interest in each image. We used default values for SIFT parameters, and calculated pairwise similarity measures for all images using the SIFT-based image-similarity measure described in Section 2. Images were considered linked if they had more than 4 SIFT matches between them. Note that the choice of 4 as a threshold was done empirically using properties of the network.

Network Validation and Characterization. At first, the size of the network created in the previous step may lead us to believe that we are limited in our ability to look at topological characteristics. However, even with the sampling and the size we experiment with in this paper, we can still show that the image network has characteristics found in other real networks. We compared properties of our image network with two networks known to be complex: film actors and the Internet. Table 1 shows the result of this comparison. The average path length (ℓ) is smaller than the film actors' and the Internet's, suggesting that a virus spread in our network could result in an epidemic faster than in film actors network. Also, The exponent of the power-law degree distribution (λ) indicates that our image network is scale free and since it is larger than the other two networks, targeted attacks (i.e., annotations) are likely to be more effective than random attacks.

Table 1. Characteristics of our network compared to Internet [5] and Film Actors [1] networks

Characteristic	Images	Film Actors	Internet
Nodes (n)	516	449,913	10,697
Links (m)	4,417	25,516,482	31,992
Mean degree (z)	8.56	113.43	5.98
Network diameter (d)	7	–	–
Degree Power Law (λ)	3.932	2.3	2.5
Average Clustering Coefficient (C)	0.207	0.20	0.035
Average Path Length (ℓ)	2.16	3.48	3.31

Target Annotation and Label Propagation. We performed experiments on all the approaches described earlier. For label propagation, we used both uniform and non-uniform propagations coupled with one of the three targeting approaches: random, HDT, and HRT. Figure 3 shows the results. All the graphs on the top row refers to uniform propagation using $\beta_{ij} = 0.5, \forall i, j$. On the bottom row, we show results for non-uniform propagation where the value of β_{ij} is given by the values of the edges, $w(e_{ij})$ in the image network. Here, both uniform and non-uniform propagations had similar performance at the end of the execution. However, non-uniform propagation converged faster to the a high number of correct annotations.

---

[1] http://pascallin.ecs.soton.ac.uk/challenges/VOC/databases.html

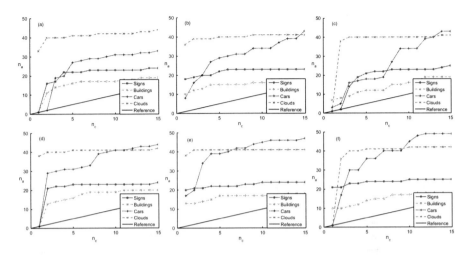

Fig. 3. Targeted labeling. n_a is the number of (manual) annotations and n_c is the number of correct labeled images. Plots show the results for different labels. Top row: use uniform propagation. Bottom row: non-uniform propagation. (a,d) show random-targeted annotation. (d,e) are highest-degree-targeted annotation, and (c,f) are page-rank-targeted annotation.

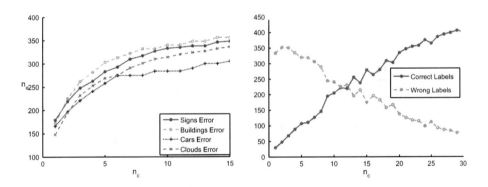

Fig. 4. False positive rates. Left: the number of wrongly labelled images according to the ground-truth. Right: if multiple labels are allowed the issue of false positives rapidly disappears.

The targeting process was performed on an unlabeled dataset. At every step of the execution, one manual label was introduced in the network. We executed it for 15 steps to verify the effect of different forms of targeting. In all the figures, a solid line is used for as a reference. That case represents the fact that 15 annotations were done and no propagation was in place.

We have used five labels and experimented on the same network but having one label propagating at a time. This way we could study the effects of each label propagation separately. As expected, different labels had widely different propagation behavior. This takes place because the propagation of a label depends on the images themselves. We presented different labels because averaging them would not be meaningful. For instance, if a network has 80% of cars and 20% of houses, we would not expect the label "house" to spread to as many images as the label "car".

The effect of different targeting algorithms is shown in each row of Figure 3. In the first column (a,d) we have random targeting, in the second column (b,e) we have HDT, and in the bottom column (c,f) we have the results for HRT. We can observe that we have good results by random targeting and even better results by targeting images with specific properties. HRT performed better than the other two methods particularly at the end of the execution, which indicates ranked-based approaches appear to perform better in partially labeled networks.

5   Conclusion and Future Work

We presented a metadata-propagation mechanism for image databases. The main idea underlying our method is to represent an image database as a network of images connected by a part-based appearance-similarity measure. We then proposed a metadata propagation method based on an immunization model. Our experiments demonstrate the feasibility of our approach. Future investigation directions include improving the similarity metric by incorporating information such as object shape, color, and texture. Another interesting direction is to explore the concept of fading [10] (i.e., labels should become weaker as they propagate in the network). Here, we could look into the case in which fading values represent a probability of a label being propagated, creating a label-specific propagation behavior.

A final issue that deserves attention relates to false positives (i.e., incorrectly labelled images). This problem is not handled in this paper. However, it is worth noticing that although we have a high rate of incorrect labelled images (see Figure 4(left)) the problem arises because in our experiments we are propagating one label at a time and each image only accepts one label (i.e., an image is either a car or not, a cloud or not, etc). Given the complexity of images in the dataset (i.e., images of buildings in clouded skies), our experiments can be seen as a proof of concept. The "incorrect" labeling is decided based on a ground-truth labeling for each image. So if an image contains a building and clouds, and we are propagating the label "cloud", our approach will assign that label to the image. However, the result in Figure 4(left) is counting that label as an error because the ground-truth says that we have a building. It is easy to see that the correct answer should actually be to label the picture as mixed of "building" and "cloud". If we allow multiple labels to propagate (Figure 4(right)), the number of wrong labels decreases rapidly which suggests again the merit of our proposal as a general concept.

References

1. Amaral, L., Scala, A., Barthelemy, M., Stanley, H.: Classes of small-world networks. Proc. of the National Academy of Sciences of the United States of America 97(21), 11149 (2000)
2. Barabási, A.-L., Albert, R.: Emergence of scaling in random networks. Science 286(5439), 509–512 (1999)
3. Brin, S., Page, L.: The anatomy of a large-scale hypertextual web search engine. Computer Networks and ISDN Systems 30(1-7), 107–117 (1998)

4. Cai, D., He, X., Li, Z., Ma, W., Wen, J.: Hierarchical clustering of www image search results using visual, textual and link information. In: Proceedings of the 12th Annual ACM International Conference on Multimedia, pp. 952–959. ACM, New York (2004)
5. Chen, Q., Chang, H., Govindan, R., Jamin, S., Shenker, S., Willinger, W.: The origin of power-laws in internet topologies revisited. In: INFOCOM, vol. 2, pp. 608–617 (2002)
6. Granovetter, M.: The strength of weak ties: A network theory revisited. Sociological Theory 1, 201–233 (1983)
7. Grauman, K., Darrell, T.: The pyramid match kernel: Efficient learning with sets of features. Journal of Machine Learning Research 8(2), 725–760 (2007)
8. Hare, J., Lewis, P.: Saliency-based models of image content and their application to auto-annotation by semantic propagation. In: Proceedings of Multimedia and the Semantic Web/European Semantic Web Conference (2005)
9. Lowe, D.: Distinctive image features from scale-invariant keypoints. International Journal of Computer Vision 60(2), 91–110 (2004)
10. Menezes, R., Wood, A.: The *fading* concept in tuple-space systems. In: ACM Symposium on Applied Computing, Dijon, France, pp. 440–444. ACM Press, New York (2006)
11. Pastor-Satorras, R., Vespignani, A.: Epidemic spreading in scale-free networks. Phys. Rev. Lett. 86(14), 3200–3203 (2001)
12. Pastor-Satorras, R., Vespignani, A.: Epidemics and immunization in scale-free networks. In: Handbook of Graphs and Networks, pp. 111–130. Wiley Press, Chichester (2003)
13. Quack, T., Ferrari, V., Leibe, B., Van Gool, L., Zurich, E., Leuven, K., Zurich, S., Oxford, U., Leuven, B.: Efficient mining of frequent and distinctive feature configurations. In: Proc. of IEEE International Conference on Computer Vision. IEEE, Los Alamitos (2007)
14. Von Ahn, L., Dabbish, L.: Labeling images with a computer game. In: Proceedings of the SIGCHI Conference on Human Factors in Computing Systems, pp. 319–326. ACM, New York (2004)
15. Watts, D.J., Strogatz, S.H.: Collective dynamics of small world networks. Nature 393, 440–442 (1998)

Epidemics in Anisotropic Networks of Roots

T.P. Handford[1], F.J. Perez-Reche[1], S.N. Taraskin[2], L. da Fontoura Costa[3,4],
M. Miazaki[3], F.M. Neri[5], and C.A. Gilligan[5]

[1] Department of Chemistry, University of Cambridge, Cambridge, UK
tph32@cam.ac.uk, p.perezreche@abertay.ac.uk
[2] St. Catharine's College and Department of Chemistry,
University of Cambridge, Cambridge, UK
snt1000@cam.ac.uk
[3] Institute of Physics at São Carlos, University of São Paulo, PO Box 369,
Postal Code 13560-970, São Carlos, São Paulo, Brazil
{luciano,mauro}@ursa.ifsc.usp.br
[4] National Institute of Science and Technology for Complex Systems, Brazil
[5] Department of Plant Sciences, University of Cambridge, Cambridge, UK
fmn22@cam.ac.uk, cag1@cam.ac.uk

Abstract. The spread of epidemics is studied in an anisotropic network of three-dimensional bean roots placed on a square lattice. In particular, the effect of global anisotropy caused for example by an external resource field is analysed. It is demonstrated that global anisotropy leads to reduced resilience to epidemic invasion as compared with a similar system of roots in the absence of external field. The origin of this effect is suggested to be correlations in transmission of infection between pairs of roots.

Keywords: epidemics, correlations, invasion.

1 Introduction

The spread of epidemics in complex networks such as biological populations of humans, animals, plants and some non-biological systems such as e.g. computer networks are under intense study amongst several scientific communities including epidemiologists, physicists, mathematicians and computer scientists [8,1]. One of the main aims in investigating epidemics is to develop and study mathematical models [9] to understand the underlying complex processes and provide effective approaches for their control. Typically the population of hosts exposed to an epidemic is split into sub-populations/classes depending on their state. In the SIR epidemiological model, the hosts can be in one of three states, susceptible (S), infected (I) and removed/recovered (R). The state of the hosts can change with time according to model-dependent rules and thus the number of hosts in each class varies with time.

One may consider SIR epidemics either in homogeneous populations [4] where the transmission probabilities of a pathogen (transmissibilities), T_{ij}, between two hosts i and j are the same for all pairs, $T_{ij} = T$, or in heterogeneous populations with random values for transmissibilities [7,6]. It has been found that in the absence of correlations in T_{ij} transmission of infection in such systems can be described by an equivalent effective

L. da F. Costa et al. (Eds.): CompleNet 2010, CCIS 116, pp. 146–153, 2011.

mean-field homogeneous system [12,11]. However, the root systems studied recently in Ref. [5] exhibited significant short-range negative correlations in transmissibilities. Such correlations are due to the fact that the root systems are typically anisotropic with respect to rotations about the vertical axis which passes through the origin of the tap root (the location where the seed was originally placed). This means that the overlap with another root system depends on direction, illustrating the nature of negative correlations: if the overlap is greater in a given direction, then it is likely to be smaller in the opposite direction. The presence of correlations breaks the mean-field description for epidemics and makes the correlated systems more resilient to invasion [10].

Here, we extend the study of SIR epidemics in a population of realistic root systems placed on a square lattice started in Ref. [5]. In the previous work, the realistic, locally anisotropic root systems were placed on lattice nodes without any predominant orientation, i.e. without any global anisotropy in root orientation. Such a design could correspond to plants in a field without any resource gradient (e.g. temperature or nutrient gradient). However, an external resource field (gradient in resources) might be present in the field and thus could cause global anisotropy in root orientations, i.e. the root systems might predominantly grow in a certain direction. This may arise in response to water, nutrient or temperature gradients within agricultural crops. The aim of this paper is to study the effect of global anisotropy on epidemics in a population of root systems. Our main finding is that global anisotropy influences the invasion probability and makes the population typically less resilient to SIR epidemics.

2   Model

2.1   SIR Process

We define the SIR process on a lattice in the following way [4]. The hosts which can be susceptible (S), infected (I) or recovered (R) are placed on the nodes of a regular lattice. An epidemic in a population of all susceptible hosts starts from infection of a single host, i (typically, in the middle of the lattice) at the initial moment of time, $t = 0$. This infected node remains in such a state during time τ_i and then it becomes recovered until the end of the epidemic, i.e. it cannot be reinfected. When the node is infected it can pass the infection to any of its nearest neighbours, e.g. node j, according to a homogeneous Poisson process with constant rate β_{ij}, so that the probability of transmitting infection from node i to node j (transmissibility) is given by,

$$T_{ij} = 1 - e^{-\beta_{ij}\tau_i} . \tag{1}$$

The process (SIR epidemic) continues according to the same dynamical rules.

Depending on the parameters of the model there are two regimes for SIR epidemics. If all $T_{ij} \ll 1$, then infection can hardly spread away from the initially infected host and the epidemic dies out leaving a macroscopically small number of recovered (previously infected) hosts. This defines the non-invasive regime for epidemics. In the opposite case, all $T_{ij} \simeq 1$, and the epidemic lasts forever in an infinite system or, in a finite system, terminates leaving a macroscopically large number (finite part of the population) of recovered hosts. This defines the invasive regime for epidemics. For intermediate values of transmissibilities, which can be conveniently characterized by mean transmissibility, $\langle T \rangle$, there is a critical regime at $\langle T \rangle = T_c$ separating the invasive and non-invasive regimes where T_c is the critical (threshold) value of transmissibility, being an important characteristic for epidemics.

The known relationship between the final state of the SIR process and bond percolation [4] allows one to identify the invasion threshold for SIR epidemics exhibiting non-correlated values of transmissibility with known bond percolation thresholds, $p_{c,\text{bond}}$ for the same topologies [12,11], i.e.

$$\langle T \rangle = T_c = p_{c,\text{bond}} , \tag{2}$$

where $p_{c,\text{bond}} = 1/2$ on the square lattice [13]. Eq. (2) holds only for independent values of T_{ij}. In the presence of positive correlations between transmissibilities, Eq. (2) is no longer valid and only the bounds are known for T_c [7,2], $p_{c,\text{bond}} \leq T_c \leq p_{c,\text{site}}$, where $p_{c,\text{site}} \simeq 0.593$ is the critical value for site percolation on the square lattice [13]. For negative correlations which arise from topologically complex host shapes [5], even the bounds for critical transmissibility are not known. This makes the present study non-trivial and quite intriguing.

2.2 Root Systems as Hosts

The hosts of the SIR model in our approach are realistic bean roots. These roots were grown and analysed using the following procedure: nine bean seeds were allowed to germinate for 2 days on cotton, and then transferred to an aquarium to grow hydroponically over 4 days, before being set into paraffin blocks which were then sliced into 0.1 mm thick layers. The layers were digitally scanned and converted to images characterized by pixels of linear size equal to 0.1 mm. Three-dimensional images of the primary (tap) root and first-order lateral roots, together comprising each root system, were thus produced (see a typical image in Fig. 1(b)).

The in silico root systems were placed vertically on the nodes of a square lattice of size $L \times L$ with lattice spacing a in such a way that the origin of the root (the position of the seed from which the tap root grows) coincided with the node (see Fig. 1(a)). The transmissibilities between nearest-neighbour roots are defined through their density-overlap integrals, J_{ij},

$$T_{ij} = 1 - e^{-\beta_{ij}\tau_i} = 1 - e^{-kJ_{ij}\tau} , \tag{3}$$

where, for simplicity, we assume host independent recovery times $\tau_i = \tau$, defined as a unit of time for the process, i.e. $\tau = 1$.

The overlap integrals are defined through the microscopic root densities (number of voxels per unit volume), $\rho_i(\mathbf{r}) = \sum_{n=1}^{N_i} \delta(\mathbf{r} - \mathbf{r}_{ni})$, where $\mathbf{r}_{ni}$ is the position vector of the n-th voxel in the image of root i and N_i is the total number of voxels for this image, in the following way,

$$J_{ij} = V_{\text{voxel}} \int d\mathbf{r}\rho_i(\mathbf{r})\rho_j(\mathbf{r}) , \tag{4}$$

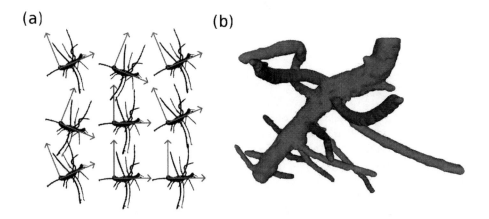

Fig. 1. Panel (a): projections onto a horizontal plane of one of the root systems (root 1 shown in panel (b)) placed on a square lattice with spacing a. This panel represents a typical pattern of the lattice model in which the root systems at each node are rotated about the vertical axis passing through the root origin (seed location) by random angle φ distributed according to Eq. (5).

with V_{voxel} (three-dimensional pixels). Technically, the overlap integral was evaluated using Gaussian broadening of δ–functions of typical width $\sigma = 1.5$ px. Such broadening accounts for the diffusive spread of a pathogen into the medium surrounding the root [3]. The coefficient of proportionality, k, between the infection rate and the overlap integral is called the infection efficiency and has a meaning of transmission rate per overlapping voxel. It is assumed to be the same for all overlaps.

In order to study the effect of an external resource field, we assume that the field is applied along the horizontal x-axis (see Fig. 1). This field forces the roots to grow predominantly along this direction. The roots are locally anisotropic with respect to rotations about the vertical axis passing through the lattice node. This means that the root density depends on the polar angle φ (within y-x plane) and attains a maximum value in a certain direction. In order to mimic the effect of the external resource field we placed the same root system (one out of nine) on each node of the lattice in such a way that the maximum density was achieved for $\varphi = 0$, i.e along the x-axis, and allowed for random rotations of roots by polar angle φ according to the following probability density function,

$$f_n(\varphi) = (2n + 1 - \delta_{n,N})^{-1} \left[\sum_{k=-n}^{n} \delta(\varphi + k\varphi_0) - \delta_{n,N}\delta(\varphi - \pi) \right], \tag{5}$$

where $\varphi_0 = \pi/N$ and $0 \leq n \leq N$ (with $N, n \in \mathbb{N}$). This function describes uniformly distributed discrete rotations in the range, $\varphi \in [-\Delta\varphi, \Delta\varphi]$ with $\Delta\varphi = n\varphi_0$ and φ_0 being the elementary rotation angle which divides π into an integer number, N, of partitions. Such a form of $f_n(\varphi)$ allows different strengths of the external field to be studied. Indeed, if $\Delta\varphi = \pi$ then all the root orientations are equivalent and thus the effect of the external field is negligible. In this case, a global anisotropy for the population does not exist while the local anisotropy defined by the shape of the root is still present. In the opposite case, $\Delta\varphi = 0$, all the root systems are oriented along the x−axis so that the local anisotropies coherently induce global anisotropy in the population along the external field.

3   Results

The quantities used to characterise the spread of epidemics are the probability of invasion, P_{inv} and the critical value of the mean transmissibility, T_c. The probability of invasion in a finite system is defined as the relative number of invasive clusters in many stochastic realizations of epidemics. The invasive cluster is the final cluster of recovered hosts spanning the whole population (touching all four boundaries of the system). In an infinite system, the invasion probability vanishes below the invasion threshold and it is finite above the threshold.

The dependence of the invasion probability on mean transmissibility (invasion curves) for two representative types of root systems (called root 1 and root 4) arranged on the lattice as shown in Fig. 1(a) are presented in Fig. 2. The invasive curves numerically calculated by averaging over 10^3 realizations of epidemics for finite-size systems ($L = 200$) have a typical sigmoidal shape (see e.g. [5]) with vanishing values for

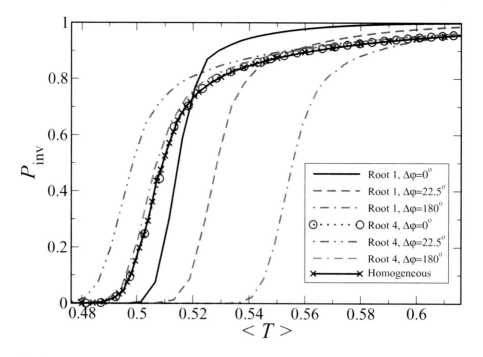

Fig. 2. Dependence of the invasion probability, P_{inv}, on the transmissibility, $\langle T \rangle$, for two types of root systems arranged on a square lattice of linear size $L = 200$ as shown in Fig. 1(a). Each curve corresponds to a different strength of external resource field determined by the parameter $\Delta\varphi$ (as marked in the legend). The solid curve marked by crosses represents a homogeneous population of root systems with all identical transmissibilities.

$\langle T \rangle < T_c$ (non-invasive regime) and finite increasing values for $\langle T \rangle > T_c$. The critical value of mean transmissibility T_c (where P_{inv} becomes close to zero) can be rigorously evaluated using finite size scaling [5].

As follows from Fig. 2, the invasion curves depend on the strength of the external resource field defined by the parameter $\Delta\varphi$. Consequently, the threshold values for mean transmissibility also depend on $\Delta\varphi$. This dependence for several representative types of root systems is shown in Fig. 3. In the case of a strong field corresponding to maximal global anisotropy, $\Delta\varphi = 0$, the SIR process can be mapped onto the anisotropic bond-percolation problem, characterized by two distinct transmissibilities T_x and T_y, corresponding to overlaps in the directions of the two horizontal axes, x and y. The threshold value for mean transmissibility, in this case, can be found exactly [14] from the equation $T_x + T_y = 1$ (which holds in the critical regime). Therefore, at criticality $\langle T \rangle = (T_x + T_y)/2 = 1/2$, coinciding with the threshold for a homogeneous system for all root systems (as can be seen from Fig. 3).

In the other extreme case, $\Delta\varphi = \pi$, when there is no influence of the external resource field, the invasion threshold is typically found to be around or above the threshold for the homogeneous system, i.e. $T_c(\Delta\varphi = \pi) \gtrsim T_{c,\mathrm{hom}} = 1/2$. However, the position of the invasion threshold below the homogeneous value cannot be excluded in principle and, in fact, we have found for a finite-size system that the invasion curve for the root system of type 4 is to the left of the homogeneous curve (cf. the double-dash-dotted line and the solid line marked by crosses in Fig. 2).

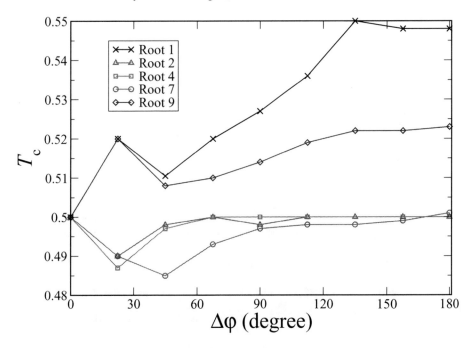

Fig. 3. Dependence of the invasion threshold T_c on the width, $\Delta\varphi$, of the distribution of local random rotations of the root systems as defined by Eq. (5) for several types of root system (as marked in the legend). The values of T_c were calculated using finite size scaling [5].

4   Discussion

Our main findings are shown in Fig. 3. The first point to be emphasised is that the external resource field can make the population of root systems less resilient to invasion of SIR epidemics as compared with the case without the external field. Indeed, this can be seen from the dependencies $T_c(\Delta\varphi)$ for roots 2, 4 and 7 in Fig. 3 such that $T_c < T_{c.\text{hom}} = 1/2$ for intermediate values of $\Delta\varphi$. A very strong external resource field does not reduce the threshold below the value for a homogeneous system, $T_c(\Delta\varphi = 0) = T_{c.\text{hom}}$, and only lower fields which allow for some local angular fluctuations of the root systems can be dangerous in the sense that $T_c < T_{c.\text{hom}}$. Secondly we note that a further increase in disorder for local orientations makes the system globally isotropic and results in an increase of the invasion threshold, such that $T_c(\Delta\varphi = \pi)$ is typically around or greater than $T_{c.\text{hom}}$. Interestingly there is great variation between the different roots systems studied and it would be instructive to study how the combination of different root systems would affect the result. In addition to this extension, an analytical understanding of the affect of correlations associated with a resource gradient could in principle be gained, in terms of models based on those for the case of no resource gradient presented in Ref. [5].

TPH would like to thank the UK EPSRC for financial support. FJPR, SNT, FMN and CAG thank BBSRC for funding (Grant No. BB/E017312/1). Luciano da F. Costa thanks CNPq (301303/2006-1 and 573583/2008-0) and FAPESP (05/00587-5) for sponsorship. Mauro Miazaki thanks FAPESP (07/50988-1) for his grant and Christopher Gilligan gratefully acknowledges the support of a BBSRC Professorial Fellowship. The authors thank Alexandre Cristino for suggestions regarding the growth of the seeds.

References

1. Barrat, A., Barthélemy, M., Vespignani, A.: Dynamical processes on complex networks. Cambridge University Press, Cambridge (2008)
2. Cox, J., Durrett, R.: Limit theorems for the spread of epidemics and forest fires. Stoch. Proc. Appl. 30, 171–191 (1988)
3. Gilligan, C.A.: Modelling soil-borne plant pathogens: reaction-diffusion models. Canadian Journal of Plant Pathology 17 (1995)
4. Grassberger, P.: On the critical behavior of the general epidemic process and dynamical percolation. Math. Biosc. 63, 157–172 (1983)
5. Handford, T.P., Pérez-Reche, F.J., Taraskin, S.N., da Costa, L.F., Miazaki, M., Neri, F.M., Gilligan, C.A.: Epidemics in networks of spatially correlated three-dimensional branching structures. Journal of the Society Interface 8(56), 423–434 (2011)
6. Kenah, E., Robins, J.M.: Second look at the spread of epidemics on networks. Phys. Rev. E 76, 036113 (2007)
7. Kuulasmaa, K.: The spatial general epidemic and locality dependent random graphs. J. Appl. Prob. 19, 745–758 (1982)
8. Murray, J.D.: Mathematical Biology. I. An Introduction, 3rd edn. Springer, Heidelberg (2002)
9. Nâsell, I.: Stochastic models of some endemic infections. Math. Biosci. 179, 1 (2002)
10. Neri, F.M., Pérez-Reche, F.J., Taraskin, S.N., Gilligan, C.A.: Heterogeneity in susceptible-infected-removed (SIR) epidemics on lattices. Journal of the Royal Society Interface 8(55), 201–209 (2011)

11. Pérez-Reche, F.J., Taraskin, S.N., da Costa, L.F., Neri, F.M., Gilligan, C.A.: Complexity and anisotropy in host morphology make populations less susceptible to epidemic outbreaks. Journal of the Royal Society Interface 7(48), 1083–1092 (2010)
12. Sander, L., Warren, C.P., Sokolov, I.M., Simon, C., Koopman, J.: Percolation on heterogeneous networks as a model for epidemics. Math. Biosc. 180, 293–305 (2002)
13. Stauffer, D., Aharony, A.: Introduction to Percolation Theory, 2nd edn. Taylor and Francis, London (1992)
14. Sykes, M.F., Essam, J.W.: Exact critical percolation probabilities for site and bond problems in two dimensions. J. Math. Phys. 5, 1117–1127 (1964)

Opinion Discrimination Using Complex Network Features

Diego R. Amancio, Renato Fabbri, Osvaldo N. Oliveira Jr.,
Maria G.V. Nunes, and Luciano da Fontoura Costa

University of São Paulo, São Carlos, São Paulo, Brazil
diego.amancio@usp.br, renato.fabbri@gmail.com, chu@ifsc.usp.br,
gracan@icmc.usp.br, ldfcosta@gmail.com

Abstract. Topological and dynamic features of complex networks have proven in recent years to be suitable for capturing text characteristics, with various applications in natural language processing. In this article we show that texts with positive and negative opinions can be distinguished from each other when represented as complex networks. The distinction was possible with the use of several metrics, including degrees, clustering coefficient, shortest paths, global efficiency, closeness and accessibility. The multidimensional dataset was projected into a 2-dimensional space with the principal component analysis. The distinction was quantified using machine learning algorithms, which allowed a recall of 84.4% in the automatic discrimination for the negative opinions, even without attempts to optimize the pattern recognition process.

1 Introduction

The use of statistical methods is well established for a number of natural language processing tasks, in some cases combined with a deep linguistic treatment in hybrid approaches [1]. In the light of the recent rapid development of network studies in various disciplines, interest in networks related to natural language is also growing. Lexical networks can be built based on different relationships between words, either of semantic or syntactic nature. These networks can exhibit both the small-world and free-scale features. In fact, representing text as graphs [2] has become popular with the advent of complex networks (CN) [3,4], especially after it was shown that large pieces of text generate scale-free networks [5,6]. This scale-free nature is probably the main reason why complex networks concepts are capable of capturing features of text, even in the absence of any linguistic treatment.

The topology and the dynamics of CN can be exploited in natural language processing, which has led to several contributions in the literature. For instance, metrics of CN have been used to assess the quality of written essays by high school students [2]. Furthermore, degrees, shortest paths and other metrics of CN were used to produce strategies for automatic summarization [7], whose results are among the best for methods that only employ statistics. The quality of machine translation systems can be examined using local mappings of local measures [8,9]. Other related applications include lexical resources analysis [10], human-induced words association [11], language evolution [12], and authorship recognition [13].

L. da F. Costa et al. (Eds.): CompleNet 2010, CCIS 116, pp. 154–162, 2011.

There are various ways to obtain a CN representation of text. The rationale is based on the intuitive fact that relations among synonyms, antonyms, hyponyms or hyperonyms may result in links. In this paper, we model texts as CN with each word being represented by a node and co-occurrences of words defining the edges. That is to say, each word is a node (the same word is reflected on the *same* node) and links are formed between two words if they appear next to each other in the text (see next section). Unlike traditional methods of text mining and sentiment detection of reviews [14,15], the method described here only takes into account the relationships between concepts, regardless of the semantics related to each word. Specifically, we analyze the topology of the networks in order to distinguish between texts with positive and negative opinions. Using a corpus of 290 pieces of text with half of positive opinions, we show that the network features allow one to achieve a reasonable distinction.

2 Methodology

The corpus employed in the experiments was obtained from the Brazilian newspaper Folha de São Paulo[1] over a 10-year period. Both positive and negative opinions were extracted from a section in which positive and negative opinions about a given subject (or fact) are confronted. Significantly, to generate a network we joined several pieces of text to reach a given length, for otherwise short pieces of texts do not provide meaningful statistical data for the CN analysis. Each CN used had approximately 1200 nodes.

To characterize the topology and dynamics, the following metrics were extracted from each network: degrees, clustering coefficient, shortest paths, global efficiency, closeness [16] and accessibility [17]. For each local measurement, six attributes were generated to characterize a given network: mean, standard deviation, median, first and third quartile over all the nodes of the network. Therefore, there are 31 attributes describing the networks, since we employed six local measurements and one global measurement (global efficiency). Then, the dataset was reduced into a two-dimensional plan using Principal Component Analysis and Linear Discriminant Analysis. Finally, the resulting dataset was examined with four machine learning inductors. The accuracy rate was analyzed using the 10-fold-cross-validation method. The main steps described here are discussed below in more detail.

2.1 Modeling Texts as Complex Networks

The present study models texts as complex networks considering each word of the text as a node and each co-occurrence of words as a link, ignoring the punctuation. Note that we could have considered the puntuactions so that only the words in the same sentence would have been connected. However, we believe that the approach adopted here is effective as it has been used successfully in other linguistic problems [2,8,9]. Furthermore, if the punctuation were considered there would be only slight changes in the results, since a single edge would be removed in each sentence. In order to combine words with the same canonical form but with different inflections into a single node,

---

[1] http://www.folha.com.br

Table 1. Pre-processed form of the sentence "The projection of the dataset into three dimensions is essential for very large datasets". Note that stopwords were removed and the remaining words were replaced by their canonical forms.

Original sentence	Pre-processed sentence
The projection of the dataset into three dimensions is essential for very large datasets	projection dataset three dimension be essential very large dataset

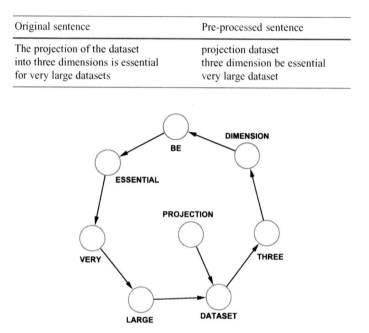

Fig. 1. Example of network generated after modeling the sentence The projection of the dataset into three dimensions is essential for very large datasets. Note that stopwords were removed and the remaining words were replaced by their respective canonical forms.

stopwords[2] were removed and the remaining words were lemmatized. Moreover, each word was labeled using the MXPost part-of-speech Tagger based on the Ratnaparkis model [18], which helps to resolve problems of ambiguity. This procedure is essential for the model, since words with the same canonical form and same meaning are grouped into a single node, while words that have the same canonical form but distinct meanings generate distinct nodes. By way of illustration, Table 1 shows the pre-processed form of the sentence "The projection of the dataset into three dimensions is essential for very large datasets" and Figure 1 shows the corresponding network for the same sentence.

2.2   Principal Component Analysis

The method referred to as Principal Components Analysis (PCA) is used in this study for data visualization and dimensionality reduction. The number of possible attributes is reduced by projecting the dataset into a two-dimensional plane in order to avoid the "curse of dimensionality" [19]. This type of projection is a statistical method that involves diagonalization of the covariance matrix of the data. In particular, strongly

---

[2] Stopwords are words with no semantic meaning such as articles and prepositions.

correlated metrics are combined to generate a set of uncorrelated variables named *principal components* [20]. Therefore, PCA seeks a linear combination of variables so that the variance of the data is concentrated along the principal components. In order to compute the principal axis of the PCA, one considers a set of observations, described by $v = \{1, 2, \ldots, Q\}$. Let M be the set of initial measurements and $\mathbf{f_v}$ be the *feature vector*, i.e., the M-dimensional vector that describes each observation. To verify the correlation between each pair of metrics i and j that describes the dataset, the *covariance* (equation 1) is computed and stored in the *covariance matrix*, defined as $C = [C(i, j)]$, with dimension $M \times M$.

$$C(i, j) = \frac{1}{Q-1} \sum_{v=1}^{Q} (f_v(i) - \mu_i)(f_v(j) - \mu_j), \tag{1}$$

$$\mu_i = \frac{1}{Q} \sum_{v=1}^{Q} f_v(i). \tag{2}$$

After defining C, the eigenvalue λ_i, $i = 1, 2, \ldots, M$; of C are computed. Upon sorting these eigenvalues in decreasing order, i.e. $\lambda_1 > \lambda_2 > \ldots \lambda_M$, the corresponding eigenvectors can be stacked to obtain the transformation matrix G [21] in equation 3. Since C is a symmetric matrix, λ_i is real (not complex) and therefore the eigenvectors are orthogonal. Consequently, G is a rotation matrix.

$$G = \begin{bmatrix} \uparrow & \uparrow & \cdots & \uparrow \\ \mathbf{v_1} & \mathbf{v_2} & \cdots & \mathbf{v_m} \\ \downarrow & \downarrow & \cdots & \downarrow \end{bmatrix}. \tag{3}$$

The new axes can be obtained from the original feature vectors $\mathbf{f}$ by the following linear transformation

$$\mathbf{g} = G\mathbf{f_v}. \tag{4}$$

Therefore, for each point i, the projection obtained is given by $g(i) = u_{11}f_i(1) + u_{12}f_i(2) + \ldots + u_{1M}f_i(M)$. In our analysis, we considered a projection into a two-dimensional space and therefore $M = 2$.

2.3   Canonical Variable Analysis

Similarly to PCA, Canonical Variable Analysis (or Linear Discriminant Analysis) is a multivariate statistical method employed to reduce dimensionality. Although this method is not used to generate attributes for the pattern recognition techniques, we compute the canonical variables to obtain a better distinction between positive and negative opinions in the dataset. Indeed, the Canonical Variable Analysis provides better distinction since the projections consider the intra-classes dispersion, which is minimized, and the inter-classes dispersion, which is maximized. To calculate the projection axes, a criterion is established to measure distances between the ζ observations. Let S be the overall dispersion of the measurements. If $\vec{x_c}$ is the set of metrics for a particular instance, and if $\langle \vec{x} \rangle$ is the average over all $\vec{x_c}$, then S is given by:

$$S = \sum_{c=1}^{\zeta} \left(\vec{x_c} - \langle \vec{x} \rangle \right) \left(\vec{x_c} - \langle \vec{x} \rangle \right)^T \tag{5}$$

Since we are dealing with two classes (C_1 = positive opinions and C_2 = negative opinions), we defined the scatter matrix S_i for each class C_i, according to equation 6, where $\langle \vec{x} \rangle_i$ is the average over all $\vec{x}$, when $\vec{x}$ is made up of instances belonging to class C_i.

$$S_i = \sum_{c \in C_i} \left(\vec{x_c} - \langle \vec{x} \rangle_i \right) \left(\vec{x_c} - \langle \vec{x} \rangle_i \right)^T \tag{6}$$

The intra-class matrix is then used to compute the dispersion inside C_1 and C_2, according to equation 7. Also, the inter-class matrix is obtained to provide the dispersion between C_1 and C_2, as shown in equation 8.

$$S_{intra} = S_1 + S_2 \tag{7}$$

$$S_{inter} = S - S_{intra} \tag{8}$$

The principal axes for the projection (called first and second variables) are then obtained by calculating the eigenvector associated with the largest eigenvalues of the matrix Λ [22], similarly to the PCA procedure, as shown in equation 9. Finally, since the data are projected in a two-dimensional space, the two principal variables are selected, corresponding to the two largest eigenvalues.

$$\Lambda = S_{intra}^{-1} S_{inter} \tag{9}$$

2.4 Pattern Recognition Methods

In order to quantify the efficiency of separation with the projection with Principal Component Analysis, we implemented machine learning algorithms. The first one is the Bayesian Decision Theory, which defines boundaries of decision according to probability density functions for each class, C_1 and C_2. This theory ensures that the classification error is minimized by assigning classifications according to the values of the likelihood function $p(C_i|\mathbf{f_v})$ with respect to $\mathbf{f_v}$ (see PCA method). In other words, $p(C_1|\mathbf{f_v})$ and $p(C_2|\mathbf{f_v})$ are compared: if the former is greater than the latter then the observation is classified as C_1, otherwise as C_2. Rewriting $p(C_i|\mathbf{f_v})$ according to the Bayes theorem (equation 10), assuming that $p(\mathbf{f_v})$ is not taken into account (because it is the same for all classes) and supposing that the a priori probabilities $p(C_1)$ and $p(C_2)$ are equal, the Bayes rule can be described as shown in equation 11.

$$p(C_i|\mathbf{f_v}) = \frac{p(\mathbf{f_v}|C_i)p(C_i)}{p(\mathbf{f_v}|C_1) + p(\mathbf{f_v}|C_2)} \tag{10}$$

$$\mathbf{f_v} \text{ is classified to } \begin{cases} C_1 \text{ if } p(\mathbf{f_v}|C_1) > p(\mathbf{f_v}|C_2) \\ C_2 \text{ otherwise} \end{cases} \tag{11}$$

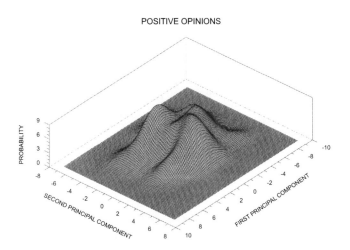

Fig. 2. Probability density found for the class of positive opinions using the Parzen windows method

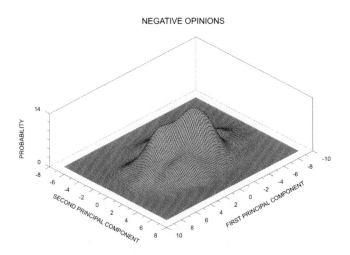

Fig. 3. Probability density found for the class of negative opinions using the Parzen windows method

The classification can be made using only the $p(\mathbf{f_v}|C_i)$ distribution. Since in most cases $p(\mathbf{f_v}|C_1)$ is not known, some estimation method is necessary. In the present work we employed the non-parametric Parzen windows method [23] to estimate both $p(\mathbf{f_v}|C_1)$ and $p(\mathbf{f_v}|C_2)$ in a two-dimensional space. This approach adds a Gaussian function at each observation point in the space, interpolatting probability densities. We employed a standard deviation equal to $\sigma_{xx} = \sigma_{yy} = 0.9$. The probabilities density obtained for each

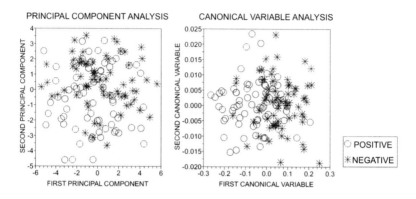

Fig. 4. Projection obtained by using Principal Component Analysis (on the left) and Canonical Variable Analysis (on the right). In both cases, there is a good distinction.

positive and negative opinions after reducing dimensionality is shown respectively in figures 2 and 3.

In addition to the Bayesian method, we used three other machine learning methods to generate inductors, namely the decision tree inductor with the C4.5 algorithm [24]; rules of decision with the RIP algorithm [25], and the Naive Bayes inductor [26]. After the training process, the efficiency of the four classifiers was evaluated using the 10-fold-cross-validation method [27].

3   Results and Discussion

The feature space employed in this work had 30 dimensions, resulting from combining six metrics extracted from the complex networks (degree, clustering coefficient, shortest paths, global efficiency, closeness and accessibility) with the five attributes generated for each measure (mean, standard deviation, median, first and third quartile). A dimension reduction was performed with PCA for pattern recognition procedures. Figure 2 shows that reasonable distinction could be achieved, particularly with the canonical variable analysis. Tables 2 and 3 summarize the accuracy for the various methods and for positive and negative opinions. The approach reached 64.79% accuracy rate with the Naive Bayes method. Further improvement is expected with the use of canonical variables for dimensionality reduction and with a systematic exploration of the feature space. Different metrics and derived attributes can be explored through automated procedures and

Table 2. % of correctly classified instances. The best classifier reached 64 % of accuracy.

Method	Accuracy Rate
Bayesian Decision	60.39 %
Decision Tree	61.27 %
Decision Rules	64.08 %
Naive Bayes	64.79 %

Table 3. Precision and Recall for positive and negative opinions. Note the high recall obtained for the negative opinions when the decision tree inductor was employed to generate the classifier.

Method	Precision Pos.	Recall Pos.	Precision Neg.	Recall Neg.
Bayesian Decision	70.2 %	49.35 %	77.3 %	71.42 %
Decision Tree	71.1 %	38.0 %	57.7 %	84.5 %
Decision Rules	70.0 %	49.3 %	60.9 %	78.9 %
Naive Bayes	70.6 %	50.7 %	61.5 %	78.9 %

should be the subject of further work. It should be highlighted that the negative opinions recall reached 84.4%, which indicates that the method is particularly useful in case one wants to have most negative opinions labeled as so. For positive opinions, the highest precision reached 71.1% with the decision tree algorithm. In Table 2, the accuracy rates are given for the four methods, showing the best accuracy rate for Naive Bayes. Even so, in Table 3 it is the decicion tree algorithm that delivers the best results, with the highest precision for positive opinions and highest recall for negative opinions.

4   Conclusion

Modeling texts as complex networks seem to have great potential for the task of distinguishing opinions. Indeed, the topological features of complex networks characterized by a set of seven measures provided a good visual distinction, as shown in the two projection methods. The efficiency was also confirmed by evaluating the accuracy rate of four machine learning inductors. Even without combining the characteristics of networks with traditional techniques, we achieved an accuracy rate of 64 %, which is reasonable. Surprisingly, we also found that the method developed here is especially good at identifying negative opinions, since the recall achieved was 84.5 %. Because of these encouraging results we will investigate in further works the effect of combining traditional methods for polarity identification with complex networks methods, in order to better capture subtleties in the texts. In addition, it is worth noticing that traditional methods may be complementary to the metrics we used in this work.

Acknowledgement. Luciano da F. Costa thanks FAPESP (05/00587-5) and CNPq (301303/06-1 and 573583/2008-0) for the financial support. Diego R. Amancio is grateful to FAPESP sponsorship (09/02941-1) and Renato Fabbri is grateful to CAPES sponsorship. We are also thankful to Dr. Oto Araujo Vale for supplying the corpus.

References

1. Manning, C.D., Schuetze, H.: Foundations of Statistical Natural Language Processing. The MIT Press, Cambridge (1999)
2. Antiqueira, L., Nunes, M.G.V., Oliveira Jr., O.N., Costa, L.F.: Strong correlations between text quality and complex networks features. Physica A 373, 811–820 (2007)
3. Newman, M.E.J.: The Structure and Function of Complex Networks. SIAM Review 45, 167–256 (2003)

4. Albert, R.Z., Barabasi, A.L.: Statistical Mechanics of Complex Networks. Rev. Modern Phys. 74, 47–97 (2002)

5. Ferrer i Cancho, R., Sole, R.V.: The small world of human language. Proceedings of the Royal Society of London B 268, 2261 (2001)

6. Barabasi, A.L.: Scale-Free Networks: a decade and beyond. Science 24, 412–413 (2009)

7. Antiqueira, L., Oliveira Jr., O.N., Costa, L.F., Nunes, M.G.V.: A Complex Network Approach to Text Summarization. Information Sciences 179(5), 584–599 (2009)

8. Amancio, D.R., Antiqueira, L., Pardo, T.A.S., Costa, L.F., Oliveira Jr., O.N., Nunes, M.G.V.: Complex networks analysis of manual and machine translations. International Journal of Modern Physics C 19(4), 583–598 (2008)

9. Amancio, D.R., Nunes, M.G.V., Oliveira Jr., O.N., Pardo, T.A.S., Antiqueira, L., da Costa, L.F.: Using metrics from complex networks to evaluate machine translation. Physica A 390(1), 131–142 (2011)

10. Sigman, M., Cecchi, G.A.: Global Organization of the Wordnet Lexicon. Proceedings of the National Academy of Sciences 99, 1742–1747 (2002)

11. Costa, L.F.: What's in a name? International Journal of Modern Physics C 15, 371–379 (2004)

12. Dorogovtsev, S.V., Mendes, J.F.F.: Evolution of networks. Advances in Physics 51, 1079–1187 (2002)

13. Antiqueira, L., Pardo, T.A.S., Nunes, M.G.V., Oliveira Jr., O.N., Costa, L.F. Some issues on complex networks for author characterization. In: Proceeedings of the Workshop in Information and Human Language Technology (2006)

14. Tang, H., Tan, S., Cheng, X.: A survey on sentiment detection of reviews. Expert Systems with Applications 36(7), 10760–10773 (2009)

15. Pennebaker, J.W., Mehl, M.R., Niederhoffer, K.G.: Psychological aspects of natural language. use: our words, our selves. Annual Review of Psychology 54, 547–577 (2003)

16. Costa, L.F., et al.: Characterization of complex networks: a survey of measurements. Advances in Physics 56, 167–242 (2007)

17. Rodrigues, F.A., Costa, L. F.: A structure dynamic approach to cortical organization: Number of paths and accessibility. Journal of Neuroscience Methods, 1–10 (2009)

18. Ratnaparki, A.: A Maximum Entropy Part-Of-Speech Tagger. In: Proceedings of the Empirical Methods in Natural Language Processing Conference (1996)

19. Bishop, C.M.: Pattern Recognition and Machine Learning. Springer, New York (2006)

20. Jolliffe, I.T.: Principal component analysis. Springer, New York (2002)

21. Costa, L.F., Cesar Jr., R.M.: Shape Analysis and Classification. CRC Press, Boca Raton (2001)

22. McLachlan, G.J.: Discriminant Analysis and Statistical Pattern Recognition. Wiley, Chichester (2004)

23. Duda, R.O., Hart, P.E., Stork, D.G.: Pattern Classification. John Wiley and Sons Inc., Chichester (2001)

24. Quinlan, R.: C4.5: Programs for Machine Learning. Morgan Kaufmann Publishers, San Francisco (1993)

25. Cohen, W.W.: Fast Effective Rule Induction. In: 12 International Converence on Machine Learning, pp. 115–223 (1995)

26. John, G.H., Langley, P.: Estimating Continuous Distribution in Bayesian Classifiers. In: 11th Conference on Uncertainty in Artificial Intelligence, pp. 338–345 (1995)

27. Kohavi, R.: A study of cross-validation and bootstrap for accuracy estimation and model selection. In: Proceedings of the Fourteenth International Joint Conference on Artificial Intelligence, vol. 12, pp. 1137–1143 (1995)

Community Structure in the Multi-network of International Trade*

Matteo Barigozzi[1], Giorgio Fagiolo[2], and Giuseppe Mangioni[3]

[1] ECARES - Université libre de Bruxelles, 50 Avenue F.D. Roosevelt CP 114,
B-1050 Brussels, Belgium
Tel.: +32 (0)2 650 33 75
matteo.barigozzi@ulb.ac.be
[2] Sant'Anna School of Advanced Studies, Laboratory of Economics and
Management, Piazza Martiri della Libertà 33, I-56127 Pisa, Italy
Tel.: +39-050-883359; Fax: +39-050-883343
giorgio.fagiolo@sssup.it
[3] Dipartimento di Ingegneria Elettrica,
Elettronica e Informatica, University of Catania,
Viale Andrea Doria, 6, I-95125 Catania, Italy
giuseppe.mangioni@dieei.unict.it

Abstract. We study the community structure of the multi-network of commodity-specific and aggregate trade relations among world countries over the 1992-2003 period. We compare structures across products and time by means of the normalized mutual information index (NMI). We also compare them with exogenous community structures induced by geographical distances and regional trade agreements. We find that: (i) plastics and mineral fuels —and in general commodities belonging to the chemical sector— have the highest similarity with aggregate trade communities; (ii) both at aggregated and disaggregated levels, physical variables such as geographical distance are more correlated with the observed trade fluxes than regional-trade agreements.

1 Introduction

In the last years there was a surge of interest in the study of international-trade issues from a complex-network perspective [3, 4, 7, 8, 12–14, 18, 24, 27, 28]. Many contributions have indeed explored the evolution over time of the topological properties of the aggregate International Trade Network (ITN), aka the World Trade Web (WTW), defined as the graph of total import/export relationships between world countries in a given year.

More recently, a number of papers have instead begun to investigate the multi-network of trade [2, 23], where a commodity-specific approach is followed to unfold the aggregate ITN in many layers, each one representing import and export relationships between countries for a given commodity class [cf. also 16, 17].

---

* The Authors would like to thank four anonymous referees for their useful and insightful comments. All usual disclaimers apply.

L. da F. Costa et al. (Eds.): CompleNet 2010, CCIS 116, pp. 163–175, 2011.
© Springer-Verlag Berlin Heidelberg 2011

In this paper, we explore further the topological architecture of the multi-network of international trade studying, for the first time, its community structure. Identifying the community structure is crucial to understand the structural and functional properties of a network and to explore emergent behaviors. Community structure detection in complex networks has received a lot of attention in the recent past and many methods have been proposed to identify communities (see for an overview [10]). To date, only two papers have been trying to study the community structure of the ITN [25, 29]. However, they have both focused their analyses on the aggregate WTW. Here, we begin detecting the community structure characterizing the commodity-specific ITN. We employ data about 162 countries and 97 commodities (2-digit disaggregation), over the period 1992-2003, to build a sequence of multi ITNs. We firstly focus on the 14 top-traded and economically relevant commodities and we identify the community structure of each layer (i.e. groups of countries that mostly trade a given commodity). Studying the commodity-specific community structure of trade (in addition to that generated from the aggregate ITN) is important because identifying communities at the product-disaggregated level may help us to better understand what are the countries in the world that tend to trade the same group of products over time. This, in turn, can shed some light on both the input-output and supply-demand interdependencies between countries.

Secondly, we compare commodity-specific community structures with a number of properly-specified community benchmarks. These benchmarks are the community structures obtained from: (i) the aggregate ITN; (ii) the network of geographical closeness (i.e. the inverse of geographical distance) between our 162 countries; (iii) trade-related partitions obtained by detecting the community structure of the regional trade agreement (RTA) network. The main question we ask is whether commodity-specific community structures are similar to, or differ from, those detected in the benchmark networks. Comparisons are made using the normalized mutual information index (NMI), which is a measure of how close two partitions of the same set of N units are [5]. Understanding whether community structures detected at the commodity-specific level are similar to — or different from — those detected in the benchmark networks can shed further light on the topological architecture of the ITN. For example, comparing aggregate and product-specific community structures may tell us whether the community structure that we observe at the aggregate trade level can be explained by the aggregation of heterogeneous community structures or, conversely, trade community formation is not affected too much by the type of commodity traded. Along similar lines, comparing trade-induced communities with those obtained by the network of geographical closeness may help us to understand the extent to which the formation of trade communities is related to geographical distance (as a proxy of trade resistance factors, e.g. trade fees).

Our results show that: (i) plastics and mineral fuels —and in general commodities belonging to the chemical sector— have the highest similarity with the community structure induced by aggregated trade; (ii) both at aggregated and disaggregated levels, physical variables such as geographical distance explain more the observed trade fluxes than RTAs. In other words, being geographically close seem to matter more than having signed a RTA to explain why groups of countries entertain more tightly interrelated trade exchanges.

The rest of the paper is organized as follows. Section 2 describes the databases that we employ in our exercises. Section 3 introduces the community detection method we used in this paper. Section 4 discusses our main results. Concluding remarks are in Section 5.

2   Data and Definitions

Our bilateral trade flows data come from the United Nations Commodity Trade Database (UN-COMTRADE; see http://comtrade.un.org/). We build a balanced panel of $N = 162$ countries for which we have commodity-specific imports and exports flows from 1992 to 2003 ($T = 12$ years) in current U.S. dollars. Trade flows are reported for $C = 97$ (2-digit) different commodities, classified according to the Harmonized System 1996 (HS1996; http://www.wcoomd.org/)[1]. The choice of a 2-digit breakdown of data may be considered insufficient to clearly identify homogeneous product lines, but it has been made because in the HS classification system there is not a unique way to further disaggregate flows by commodities at a higher number of digits. Notice, however, that network analyses often face a trade off between the need for a finer disaggregation and the very possibility to obtain connected graphs: typically, as soon as 3 or 4 digit data are considered, the resulting graphs easily become not connected, with the size of the largest connected component quickly decreasing.

We employ the database to build a time sequence of weighted, directed multi-networks of trade where the N nodes are world countries and directed links represent the value of exports of a given commodity in each year or wave $t = 1992, \ldots, 2003$. As a result, we have a time sequence of T multi-networks of international trade, each characterized by C layers (or links of C different colors). Each layer $c = 1, \ldots, C$ represents exports between countries for commodity c and can be characterized by a $N \times N$ weight matrix X_t^c. Its generic entry $x_{ij,t}^c$ corresponds to the value of exports of commodity c from country i to country j in year t. We consider directed networks, therefore in general $x_{ij,t}^c \neq x_{ji,t}^c$. The aggregate weighted, directed ITN is obtained by simply summing up all commodity-specific layers. The entries of its weight matrices X_t reads:

$$x_{ij,t} = \sum_{c=1}^{C} x_{ij,t}^c. \tag{1}$$

---

[1] Since, as always happens in trade data, exports from country i to country j are reported twice (according to the reporting country — importer or exporter) and sometimes the two figures do not match, we follow Ref. [9] and only employ import flows. For the sake of exposition, however, we follow the flow of goods and we treat imports from j to i as exports from i to j.

For the sake of exposition, we shall focus on the most important commodity networks. Table 1 shows the ten most-traded commodities in 2003, ranked according to the total value of trade. Notice that they account, together, for 56% of total world trade and that the 10 most-traded commodities feature also the highest values of trade-value per link (i.e. ratio between total trade and total number of links in the commodity-specific network). In addition to those trade-relevant 10 commodities, we shall also focus on other 4 classes (cereals, cotton, coffee/tea and arms), which are less traded but more relevant in economics terms. The 14 commodities considered account together for 57% of world trade in 2003.

We also employ data about RTAs between world countries taken from the World Trade Organization (WTO) website[2]. We build a weighted undirected network with weight matrix $M_t = \{m_{ij,t}\}$ where nodes are countries and a link is weighted according to the number $m_{ij,t}$ of RTAs – free, multilateral and/or bilateral – in place between the two countries i and j at year t [cf. also 25].

Finally, we build a geographically-related weighted undirected network with weights $s_{ij} = d_{ij}^{-1}$, where d_{ij} is the geographical distance between the main (most populated) cities of country i and country j[3]. We employ the resulting matrix $S = \{s_{ij}\}$ as a weighted undirected network of geographical closeness between countries.

3   Community Detection

It has been observed that many real networks exhibit a concentration of links within a special groups of nodes called communities (or clusters or modules). Such a structural property of a network has also been linked to the presence of sub-modules whose nodes have some functional property in common. Therefore, the detection of the community structure of a given network could help to discover some hidden feature.

Despite the intuitive concept of community, a precise definition of what a community is represents a challenging issue (see [10]). In this paper we adopt the well known formulation given in [21]: a subgraph is a community if the number of links (or, more generally, the intensity of interactions) among nodes in the subgraph is higher than what would be expected in an equivalent network with links (and intensities) placed at random. This definition implies the choice of a so–called "null model", i.e. a model of network to which any other network can be compared in order to assert the existence of any degree of modularity. The most used null model is a random network with the same number of nodes, the same number of links and the same degree distribution as in the original network, but with links among nodes randomly placed. Based on these concepts, a function called modularity that gives a measure of the quality of a given network partition into communities has been introduced in [21]. The modularity function has been further extended in [1] to the case of weighted directed networks as reported in

---
[2] See http://www.wto.org/
[3] Data are available at the URL: http://www.cepii.fr/

the following:

$$Q = \frac{1}{W} \sum_{ij} \left[w_{ij} - \frac{w_i^{out} w_j^{in}}{W} \right] \delta_{c_i, cj} \tag{2}$$

where w_{ij} is the weight of the link between i and j, $w_i^{out} = \sum_j w_{ij}$ and $w_j^{in} = \sum_i w_{ij}$ are respectively the output and input strengths of nodes i and j, $W = \sum_i \sum_j w_{ij}$ is the total strengths of the network and δ_{c_i, c_j} is 1 if nodes i and j are in the same community and 0 otherwise.

In this paper communities are uncovered by optimising the modularity function in equation (2). The optimisation of Q is performed by using a tabu search algorithm [15].

As discussed in Section 1, one of the contribution of this paper is to compare commodity–specific community structures with a proper number of community benchmarks (as detailed in the next Section). To compare community partitions we use the *normalised mutual information* (NMI) measure, as introduced in [5]. To define the NMI index, we need to introduce the confusion matrix. Given two community partitions $\mathcal{P}_A$ and $\mathcal{P}_B$, the confusion matrix $\mathcal{N}$ is defined as a matrix whose N_{ij}-th element is the number of nodes in the community i of the partition $\mathcal{P}_A$ that appear in the community j of the partition $\mathcal{P}_B$. The NMI is defined as:

$$NMI(\mathcal{P}_A, \mathcal{P}_B) = \frac{-2 \sum_{i=1}^{C_A} \sum_{j=1}^{C_B} N_{ij} log \left(\frac{N_{ij} N}{N_{i.} N_{.j}} \right)}{\sum_{i=1}^{C_A} N_{i.} log \left(\frac{N_{i.}}{N} \right) + \sum_{j=1}^{C_B} N_{.j} log \left(\frac{N_{.j}}{N} \right)} \tag{3}$$

where C_A and C_B are respectively the number of communities in $\mathcal{P}_A$ and $\mathcal{P}_B$, $N_{i.} = \sum_j N_{ij}$, $N_{.j} = \sum_i N_{ij}$ and $N = \sum_i \sum_j N_{ij}$. The NMI index is equal to 1 if $\mathcal{P}_A$ and $\mathcal{P}_B$ are identical and assumes a value of 0 if the two partitions are independent.

4   Results

Our results are based on the values of the NMI index across commodities and time. More specifically, we detect the community structure on our 14 commodity-specific trade networks in every year from 1992 to 2003. Furthermore, we obtain —as a first benchmark— the community structure for the 12 yearly aggregate trade networks with weight matrix as in eq. (1). Finally, we apply community detection on the geographical closeness matrix S, as well on the 12 RTA networks M_t. Notice that clusters identified in the network with weight matrix S represent groups of countries that are geographically close, without using exogenously-determined partitions of countries (e.g., based on continents or sub-continental breakdowns). The community structure detected in networks M_t does instead pick up clusters of countries that not only belong to free-trade or multilateral agreements (e.g. NAFTA, Mercosur, EU, etc.), but also signed additional bilateral agreements.

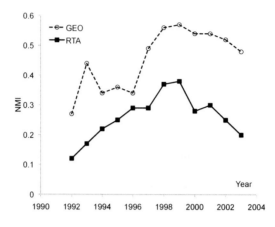

Fig. 1. NMI when comparing the community structures induced by the exogenous networks built using geographical distances (GEO) or regional trade agreements data (RTA) with the the community structures of aggregate trade

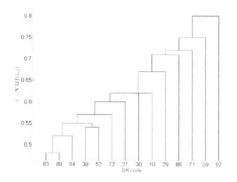

Fig. 2. Minimum spanning tree for year 2003

We first compare aggregate-trade community structure with those detected using geographical distances or RTAs (see Figure 1). We observe increasing NMIs across time until 2001 and a slight decrease afterwards. We also find more similarity between aggregate trade and geography based communities with respect to communities determined by RTAs. Thus, physical factors more than political factors seem to be more correlated with the patterns of global trade. Also, this result is more evident in the recent years after 2001. A possible explanation might be the global political crisis after 11th September 2001 that implied a slight decrease in global trade as a consequence of the wars in Iraq and Afghanistan.

We then compare aggregate trade with commodity specific communities. In general, the role of all commodities in shaping the aggregated community structure has increased over time. In particular, we observe the largest increase in NMI from 1992 to 2003 for mineral fuels ($c = 27$), plastics ($c = 39$), iron and

(a) Aggregate trade.

(b) RTA.   (c) Distances.

(d) Mineral fuels $c = 27$.   (e) Plastics $c = 39$.

Fig. 3. World maps showing communities in 2003

steel ($c = 72$), pharmaceutical products ($c = 30$), and organic chemicals ($c = 29$). Moreover, for all considered years mineral fuels and plastics have the most similar community structure to the aggregated one (see Table 2). A major role for the chemical sector appears from these results.

When comparing community structures induced by disaggregated trade data with the structures determined by geographical and RTA data, we find results similar to the aggregate case (see Table 3). Namely, it is geography and not RTAs that is more related to trade. Plastics ($c = 39$) followed by mineral fuels ($c = 27$) have the highest similarity with RTA communities (see Table 4). The same result holds when comparing trade communities with the geographical data, but in addition we notice high NMIs also for iron and steel ($c = 72$) and cotton ($c = 52$).

Our results can also be visualized using world maps. Figure 2 shows the communities obtained for 2003 data on aggregate trade, RTAs, geographical distances, and data for the most two relevant commodities: mineral fuels ($c = 27$)

Table 1. The 14 most relevant commodity classes in year 2003 in terms of total-trade value (USD), trade value per link (USD), and share of world aggregate trade

Code	Commodity	Value (USD)	Value per Link (USD)	% of Aggregate Trade
83	Nuclear reactors, boilers, machinery and me-chanical appliances; parts thereof	5.67×10^{11}	6.17×10^7	11.37%
84	Electric machinery, equipment and parts; sound equipment; television equipment	5.58×10^{11}	6.37×10^7	11.18%
27	Mineral fuels, mineral oils & products of their distillation; bitumin substances; mineral wax	4.45×10^{11}	9.91×10^7	8.92%
86	Vehicles, (not railway, tramway, rolling stock): parts and accessories	3.09×10^{11}	4.76×10^7	6.19%
89	Optical, photographic, cinematographic, mea-suring, checking, precision, medical or surgical instruments/apparatus; parts & accessories	1.78×10^{11}	2.48×10^7	3.58%
39	Plastics and articles thereof.	1.71×10^{11}	2.33×10^7	3.44%
29	Organic chemicals	1.67×10^{11}	3.29×10^7	3.35%
30	Pharmaceutical products	1.4×10^{11}	2.59×10^7	2.81%
72	Iron and steel	1.35×10^{11}	2.77×10^7	2.70%
71	Pearls, precious stones, metals, coins, etc	1.01×10^{11}	2.41×10^7	2.02%
10	Cereals	3.63×10^{10}	1.28×10^7	0.73%
52	Cotton, including yarn and woven fabric thereof	3.29×10^{10}	6.96×10^6	0.66%
9	Coffee, tea, mate & spices	1.28×10^{10}	2.56×10^6	0.26%
92	Arms and ammunition, parts and accessories thereof	4.31×10^9	2.46×10^6	0.09%
ALL	Aggregate	4.99×10^{12}	3.54×10^8	100.00%

Table 2. NMI when comparing the community structures induced by aggregate trade with commodity-specific trade

HS Code / Year	09	10	27	29	30	39	52	71	72	83	84	86	89	92
1992	0.06	0.11	0.18	0.12	0.09	0.16	0.21	0.10	0.14	0.23	0.21	0.25	0.21	0.05
1993	0.17	0.14	0.37	0.29	0.23	0.44	0.24	0.18	0.28	0.25	0.19	0.16	0.21	0.11
1994	0.11	0.15	0.23	0.31	0.18	0.32	0.28	0.13	0.18	0.35	0.35	0.22	0.40	0.12
1995	0.14	0.20	0.25	0.30	0.22	0.40	0.28	0.21	0.24	0.36	0.40	0.19	0.26	0.15
1996	0.05	0.17	0.39	0.32	0.22	0.46	0.24	0.09	0.25	0.39	0.41	0.40	0.28	0.21
1997	0.12	0.32	0.46	0.33	0.28	0.30	0.29	0.20	0.37	0.46	0.25	0.30	0.24	0.18
1998	0.18	0.25	0.54	0.33	0.34	0.39	0.27	0.18	0.42	0.42	0.29	0.37	0.30	0.16
1999	0.15	0.33	0.56	0.31	0.29	0.51	0.38	0.26	0.37	0.39	0.38	0.31	0.31	0.12
2000	0.23	0.25	0.43	0.34	0.32	0.34	0.41	0.25	0.29	0.46	0.41	0.35	0.28	0.17
2001	0.20	0.26	0.54	0.38	0.31	0.50	0.33	0.28	0.43	0.42	0.40	0.47	0.39	0.15
2002	0.24	0.33	0.54	0.29	0.28	0.35	0.39	0.21	0.35	0.28	0.33	0.28	0.28	0.16
2003	0.14	0.27	0.49	0.32	0.29	0.45	0.38	0.28	0.38	0.37	0.19	0.25	0.26	0.12

Table 3. NMI when comparing the community structures induced by geographical distances with commodity-specific trade

HS Code Year	09	10	27	29	30	39	52	71	72	83	84	86	89	92
1992	0.24	0.26	0.41	0.31	0.29	0.37	0.30	0.21	0.30	0.25	0.24	0.29	0.26	0.21
1993	0.24	0.28	0.38	0.37	0.38	0.43	0.32	0.25	0.42	0.31	0.24	0.28	0.21	0.22
1994	0.23	0.38	0.47	0.27	0.34	0.43	0.30	0.21	0.36	0.29	0.28	0.38	0.27	0.23
1995	0.26	0.41	0.52	0.30	0.33	0.54	0.27	0.31	0.43	0.34	0.31	0.31	0.31	0.30
1996	0.18	0.32	0.43	0.26	0.31	0.38	0.32	0.22	0.43	0.30	0.34	0.32	0.33	0.28
1997	0.23	0.42	0.48	0.34	0.31	0.42	0.40	0.27	0.52	0.43	0.31	0.31	0.27	0.26
1998	0.21	0.37	0.52	0.39	0.39	0.43	0.35	0.22	0.41	0.41	0.31	0.35	0.31	0.23
1999	0.20	0.43	0.56	0.30	0.29	0.47	0.48	0.32	0.44	0.42	0.44	0.26	0.36	0.21
2000	0.30	0.36	0.45	0.40	0.40	0.44	0.48	0.33	0.37	0.43	0.43	0.33	0.30	0.21
2001	0.26	0.39	0.52	0.33	0.36	0.53	0.44	0.35	0.46	0.38	0.45	0.38	0.41	0.17
2002	0.27	0.41	0.53	0.31	0.32	0.47	0.44	0.25	0.47	0.30	0.35	0.33	0.32	0.21
2003	0.20	0.39	0.50	0.33	0.37	0.57	0.44	0.29	0.52	0.36	0.27	0.33	0.32	0.19

Table 4. NMI when comparing the community structures induced by regional trade agreements (RTA) with commodity-specific trade

HS Code Year	09	10	27	29	30	39	52	71	72	83	84	86	89	92
1992	0.15	0.17	0.20	0.13	0.20	0.17	0.22	0.12	0.19	0.16	0.15	0.14	0.13	0.15
1993	0.19	0.19	0.21	0.13	0.22	0.15	0.24	0.13	0.23	0.19	0.14	0.15	0.12	0.15
1994	0.16	0.27	0.24	0.21	0.26	0.21	0.19	0.14	0.26	0.16	0.20	0.19	0.16	0.17
1995	0.18	0.29	0.37	0.18	0.32	0.38	0.23	0.18	0.31	0.30	0.27	0.25	0.24	0.20
1996	0.11	0.28	0.33	0.24	0.27	0.31	0.26	0.16	0.26	0.25	0.30	0.29	0.21	0.23
1997	0.13	0.33	0.30	0.27	0.24	0.26	0.24	0.15	0.22	0.28	0.17	0.23	0.17	0.14
1998	0.27	0.31	0.36	0.22	0.36	0.42	0.28	0.14	0.27	0.32	0.25	0.30	0.20	0.21
1999	0.19	0.33	0.37	0.20	0.32	0.41	0.30	0.23	0.31	0.24	0.24	0.25	0.22	0.21
2000	0.25	0.29	0.35	0.22	0.25	0.33	0.33	0.22	0.26	0.26	0.30	0.24	0.15	0.17
2001	0.20	0.25	0.31	0.18	0.27	0.34	0.27	0.21	0.28	0.29	0.26	0.25	0.22	0.17
2002	0.20	0.30	0.30	0.19	0.25	0.34	0.31	0.20	0.29	0.17	0.23	0.25	0.22	0.19
2003	0.16	0.24	0.32	0.20	0.29	0.33	0.24	0.16	0.24	0.25	0.21	0.26	0.19	0.18

and plastics ($c = 39$). We often observe the same kind of communities: a cluster with America, a cluster with China, India, Indocina and Oceania, a cluster with Europe that might contain also Russia, finally Africa and Middle East are divided among the other groups.

Finally, starting from the similarity between commodity i and commodity j, $\mathrm{NMI}(i,j)$ we can define a distance between two community structures as $1 - \mathrm{NMI}(i,j)$. Using this measure, we can build a minimum spanning tree as in [19] in order to classify the commodities. The result for 2003 is in Figure 3 and show how commodities related to science- or technology-based industries (nuclear reactors, optics, electrics) are the most similar in terms of their community structures, whereas arms appear to be the more dissimilar one.

5   Concluding Remarks

In this paper, we have provided a first exploratory study of the community structure of product-specific trade networks from 1992-2003. After recovering the optimal partition of countries, we have compared commodity specific communities with the aggregate trade community.

A major role for the chemical sector (mineral fuels, plastics, chemicals and pharmaceuticals) appears from this first set of results. When comparing trade communities with exogenous communities generated by geographical distances or RTAs, we find that being geographically close seem to matter more than having signed a RTA to explain why groups of countries entertain more tightly interrelated trade exchanges.

These results contribute to the discussion related to the impact of international agreements on world trade and seem to go in the direction of Ref. [26], which shows that there is no evidence that the WTO has increased world trade. Of course, our findings are only partial as they only check for unconditional effects (i.e. one does not address the residual effects of RTAs once geography is controlled for). In order to make our statements more robust, one may follow Ref. [25] and compare communities observed in trade data with those detected in the network built with the predictions of a standard gravity model [see for example 6] .

The robustness of our findings should be checked against a number of possible problems. For example, it is well-known that modularity-based community detection suffers from many possible biases [for example, resolution limits, see Ref. 11]. Therefore, community detection algorithms based on alternative criteria (e.g., information theory) may be employed. Similarly, one may consider to apply algorithms allowing for overlapping communities [22].

Finally, another point that deserves further analysis is the detection of community structures across commodity-specific layers. In the paper, we have analyzed independently the most important 14 layers. This allows one to identify groups of countries that trade the same commodity among them. From an economic point of view, this signals strong interdependencies but does not convey any insights on the input-output structure of the cluster. For example, there might be groups of countries that are linked in tightly connected chains or cycles, where a country imports from another a particular type of commodity needed as input for its peculiar industrial structure, and at the same time exports to other countries in the group another commodity that is fed into their production processes (or consumed as final good). In order to address these issues, one would like to either synthesize into a meaningful statistic all commodity-specific relationships between any two countries or apply new techniques able to detect community structures in multi graphs [20].

References

1. Arenas, A., Duch, J., Fernandez, A., Gomez, S.: Size reduction of complex networks preserving modularity. New Journal of Physics 9(6), 176 (2007)
2. Barigozzi, M., Fagiolo, G., Garlaschelli, D.: Multi-network of international trade: A commodity-specific analysis. Physical Review E 81, 046104 (2010)
3. Bhattacharya, K., Mukherjee, G., Manna, S.S.: The international trade network. In: Chatterjee, A., Chakrabarti, B.K. (eds.) Econophysics of Markets and Business Networks. Springer, Milan (2007)
4. Bhattacharya, K., Mukherjee, G., Sarämaki, J., Kaski, K., Manna, S.S.: The international trade network: Weighted network analysis and modeling. Journal of Statistical Mechanics: Theory Exp. A 2, 02002 (2008)
5. Danon, L., Diaz-Guilera, A., Duch, J., Arenas, A.: Comparing community structure identification. Journal of Statistical Mechanics: Theory and Experiment 2005(09), 09008 (2005)
6. Fagiolo, G.: The international-trade network: Gravity equations and topological properties. Journal of Economic Interaction and Coordination 5, 1–25 (2010)
7. Fagiolo, G., Schiavo, S., Reyes, J.: On the topological properties of the world trade web: A weighted network analysis. Physica A 387, 3868–3873 (2008)
8. Fagiolo, G., Schiavo, S., Reyes, J.: World-trade web: Topological properties, dynamics, and evolution. Physical Review E 79, 036115 (2009)
9. Feenstra, R.C., Lipsey, R.E., Deng, H., Ma, A.C., Mo, H.: World trade flows: 1962-2000. NBER Working Papers 11040, National Bureau of Economic Research, Inc., (2005)
10. Fortunato, S.: Community detection in graphs. Physics Reports 486, 75–174 (2010)
11. Fortunato, S., Barthélemy, M.: Resolution limit in community detection. Proceedings of the National Academy of Sciences 104(1), 36–41 (2007)
12. Garlaschelli, D., Di Matteo, T., Aste, T., Caldarelli, G., Loffredo, M.I.: Interplay between topology and dynamics in the world trade web. The European Physical Journal B 57, 1434–6028 (2007)
13. Garlaschelli, D., Loffredo, M.I.: Fitness-dependent topological properties of the world trade web. Physical Review Letters 93, 188701 (2004)
14. Garlaschelli, D., Loffredo, M.I.: Structure and evolution of the world trade network. Physica A 355, 138–144 (2005)
15. Glover, F., Laguna, M.: Tabu Search. Kluwer Academic Publishers, Dordrecht (1998)
16. Hidalgo, C.A., Klinger, B., Barabási, A.L., Hausmann, R.: The product space conditions the development of nations. Science 317(5837), 482–487 (2007)
17. Hidalgo, C.A., Hausmann, R.: The building blocks of economic complexity. Proceedings of the National Academy of Sciences 106(26), 10570–10575 (2009)

18. Li, X., Jin, Y.Y., Chen, G.: Complexity and synchronization of the world trade web. Physica A: Statistical Mechanics and its Applications 328, 287–296 (2003)
19. Mantegna, R.N.: Hierarchical structure in financial markets. The European Physical Journal B 11, 193–197 (1999)
20. Mucha, P.J., Richardson, T., Macon, K., Porter, M.A., Onnela, J.-P.: Community structure in time-dependent, multiscale, and multiplex networks. Science 328(5980), 876–878 (2010)
21. Newman, M.E.J., Girvan, M.: Finding and evaluating community structure in networks. Phys. Rev. E 69(2), 026113 (2004)
22. Nicosia, V., Mangioni, G., Carchiolo, V., Malgeri, M.: Extending the definition of modularity to directed graphs with overlapping communities. Journal of Statistical Mechanics: Theory and Experiment, An IOP and SISSA Journal, vol. 2009, p. P03024, ISSN: 1742-5468, doi:10.1088/1742-5468/2009/03/P03024
23. Reichardt, J., White, D.R.: Role models for complex networks. The European Physical Journal B 60, 217–224 (2007)
24. Reyes, J., Schiavo, S., Fagiolo, G.: Assessing the evolution of international economic integration using random-walk betweenness centrality: The cases of East Asia and Latin America. Advances in Complex Systems 11, 685–702 (2008)
25. Reyes, J., Wooster, R.B., Shirrel, S.: Regional trade agreements and the pattern of trade: A networks approach. Working paper, SSRN Working Paper Series (2009)
26. Rose, A.K.: Do we really know that the WTO increases trade? American Economic Review 94(1), 98–114 (2004)
27. Serrano, A., Boguñá, M.: Topology of the world trade web. Physical Review E 68, 015101(R) (2003)
28. Serrano, A., Boguñá, M., Vespignani, A.: Patterns of dominant flows in the world trade web. Journal of Economic Interaction and Coordination 2, 111–124 (2007)
29. Tzekina, I., Danthi, K., Rockmore, D.: Evolution of community structure in the world trade web. The European Physical Journal B - Condensed Matter 63, 541–545 (2008)

Uncovering Overlapping Community Structure

Qinna Wang and Eric Fleury

ENS de Lyon, 46 allée d'Italie, 69364 LyonCedex 07
Qinna.Wang@ens-lyon.fr, Eric.Fleury@inria.fr

Abstract. Overlapping community structure has attracted much interest in recent years since Palla et al. proposed the k-clique percolation algorithm for community detection and pointed out that the overlapping community structure is
more reasonable to capture the topology of networks. Despite many efforts to
detect overlapping communities, the overlapping community problem is still a
great challenge in complex networks. Here we introduce an approach to identify
overlapping community structure based on an efficient partition algorithm. In our
method, communities are formed by adding peripheral nodes to cores. Therefore,
communities are allowed to overlap. We show experimental studies on synthetic
networks to demonstrate that our method has excellent performances in community detection.

1 Introduction

Complex networks have attracted a lot of attention and helped in a better understanding
of the properties and the characteristics of several network systems: social [2], technological [7], biological [24], etc. Most real networks cited above have an intrinsic
structure and one may find parts of the network, within which nodes are more highly
connected to each other than to the rest of the network. Such highly connected zones
are usually called clusters, modules or communities. Communities reveal the network
organization at a mesoscopic level and also provide insights in understanding structural
functions: communities of sets of Web pages dealing with the same topic in the World
Wide Web network [7], the modular structure in biological networks showing biological
functions [9], etc.

Newman and Girvan [15] introduced modularity to measure the quality of a graph
partition. The modularity of a partition is a scalar (bounded by 1). It is used to compare
the quality of several heuristics [15] and it is also used as an objective function to optimize [23,5]. Although Fortunato et al. [8] pointed out the resolution limit of modularity,
modularity is still widely accepted and a de facto quality function and modularity optimization is considered to be the most efficient community detection approach. Recently,
the so-called Louvain method [4] outperforms clearly by several degrees of magnitude,
all known modularity optimization heuristics in terms of computer time and modularity.

The main drawback of modularity optimization is that it tries to construct a partition
of the graph (each node is classified into one and only one community). However, most
real networks reveal more complex structures with overlapping communities. For example, a scientist who is a physicist may also be a mathematician. By this reasoning,
an algorithm which allows communities to overlap with others will produce a more reasonable result. Consequently, more universal and special features of networks will be

L. da F. Costa et al. (Eds.): CompleNet 2010, CCIS 116, pp. 176–186, 2011.

mined and understood. For instance, the knowledge of overlapping community structure is useful in predicting community evolution in dynamic networks.

Here we propose a new clustering approach for uncovering overlapping communities. Taking benefits of Louvain algorithm, we uncover the community structure by adding peripheral nodes to cores. Experimental results in synthetic networks demonstrate that our method has excellent performances.

2   Related Works

We review current diverse overlapping community detection approaches, such as clique-based, matrix-based and degree-based. Palla et al. [18] proposed the Clique Percolation Method (**CPM**) that defines a community, or more precisely, a k-clique-community as a union of a sequence of adjacent k-cliques (where adjacency means sharing $k-1$ nodes). However, **CPM** fails to capture the community structure of sparse networks for rare k-cliques (see [22]). Fuzzy clustering methods like the Fuzzy C-Means (**FCM**) clustering algorithm [27] embed nodes into d-dimensional Euclidean spaces and use geometric clustering to generate cluster memberships. Nonetheless, they suffer from expensive computational cost, which is $\mathscr{O}(nK^2)$ in worst case for $K-1$ eigenvectors calculation. Baumes et al. [3] proposed a density metric for clustering nodes. In this way, nodes are added into clusters if and only if their fusions improve the cluster density but low degree nodes may form many small communities.

There are also some studies on community detection related to our work. In [21], Sales-Pardo et al. applied a box-clustering method to build an hierarchical tree. The fusion of groups of nodes is based on an affinity matrix whose element is the probability of a pair of nodes classified into the same community with respect to local modularity maxima. However, this algorithm has a high computational cost which results the limitation of network size (the size of network is limited to 10,000 nodes). Li et al. [13] proposed a core-periphery clustering algorithm to merge peripheral clusters with cores. Cores are formed by merging candidate cores if their fusion generates a 4-clique, where the candidate cores are initialized with triangles. This method may produce giant communities due to clique percolation [6]. In contrast, our method is neither limited by the scale of networks nor restricted by cliques.

3   Community Detection

In this section, we introduce important definitions in our method. The core C_s, also called the seed, are embedded in communities. The goal of our method, which is a Core-Peripheral Clustering method (**CPC**), is to expand cores by optimizing a local community fitness function. Let $G = (V,E)$ be a graph with $n = |V|$ vertices and $m = |E|$ edges.

Definition 1 (disjoint community structure). *A disjoint community structure of G is a partition of V into a set $\mathscr{P} = \{C_1,C_2,C_3,...,C_c\}$ of c nonempty subsets of V such that, by definition, every element $u \in V$ is in exactly one of these subsets: (i) The union of the elements of $\mathscr{P}$ is equal to V: $\cup \mathscr{P} = \cup_{i=1}^{c} C_i = V$ and (ii) The intersection of any two distinct elements of $\mathscr{P}$ is empty: $\forall\, 1 \leq i < j \leq c,\ C_i \cap C_j = \emptyset$.*

Definition 2 (overlapping community structure). *An* overlapping community structure *of G is a covering of the set V into a set* $\mathscr{S} = \{C_1, C_2, C_3, ..., C_c\}$ *of c nonempty
subsets of V such that the elements of $\mathscr{S}$ are covering V:* $\cup \mathscr{S} = \cup_{i=1}^{c} C_i = V$.

Definition 3 (The fitness function). *A community fitness function indicates the average internal edge density inside a subgraph C, which is given by*

$$f_C = \frac{\sum_{i \in C} k_i^{int}}{(\sum_{i \in C} k_i^{int} + \sum_{i \in C} k_i^{out})^{\alpha}}, \tag{1}$$

*where α is a tunable parameter, the internal degree k_i^{int} and external degree k_i^{out} of
node $i \in C$ are the number of edges from node i to other nodes inside of community C
or outside of community C, respectively.*

The parameter α tunes the resolution of the method. Many studies show the value of
α determines the scale of communities, and α in the range $[0.8, 1.5]$ provides good results [11]. A natural choice is $\alpha = 1$, as the ratio of the internal degree to the total degree
of the community, which also corresponds to the definition of community introduced
by Radicchi et al. [19].

3.1 Modularity

In order to quantify community structure, we introduce the modularity, which measures
the variance between the real fraction of links in communities and the expected fraction
of links in a null model (i.e. a size equivalent random graph with the same degree
distribution). Generally, a good partition should have a high value. The most popular
modularity function Q is proposed by Newman and Girvan in [15] to find and evaluate
the partition. However, it fails to evaluate a cover. We apply an extension of modularity
Q [1] to show the quality of the found community structure, which is:

$$Q = \sum_{i=1}^{c} \left[\frac{A(V_i, V_i)}{A(V, V)} - \left(\frac{A(V_i, V)}{A(V, V)} \right)^2 \right] \tag{2}$$

where $A(V_i, V_j) = \sum_{u \in V_i, v \in V_j} k_{u,v}$ and $k_{u,v}$ is the weight of edge $\{u, v\}$. The modularity
Q evaluates the difference between the probability of edges within the community and
the probability of edges incident to nodes within the community.

3.2 Greedy Expansion

In our experiments, the fitness proposed by Lancichinetti et al. [11] provided good
results on synthetic and empirical data. We treat it as a local optimization strategy for
community detection, that is, we identify each community C by adding nodes to core C_s
through a fitness optimization technique. The optimization technique can be concluded
as following:

1. Calculate the node fitness for all neighbor nodes of C, where the node fitness is
   $F_v = F_{C+v} - F_{C-v}$;

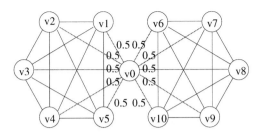

Fig. 1. A simple graph with overlapping nodes. It is composed of 2 cliques sharing node v_0. The edges in red represent $p_{ij} = 50\%$, and the edges in black represents $p_{ij} = 100\%$. Thus, the sets of nodes $\{v_1, v_2, v_3, v_4, v_5\}$ and $\{v_6, v_7, v_8, v_9, v_{10}\}$ construct strong clusters.

2. Add the node v which brings the largest positive F_v to C;
3. Recalculate the fitness of nodes in C;
4. Remove the node which has the negative fitness from C;
5. If step 4 occurs, repeat from the step 3; otherwise repeat from the step 1.

The iteration repeats until that no more movement can improve the fitness of C. In this iteration, a tunable parameter α is involved. To specify it, we tune it in $[0.8, 1.5]$ and select the value corresponding to the results similar to the ground truth. When there is no a priori knowledge of the network, we select Louvain algorithm to provide the community structure. In real networks analysis, it makes good effects.

3.3 Choices of Cores

We choose strong clusters (i.e. a set of nodes keeping stable memberships) to cores. The choice of cores is not unique. Lancichinetti et al. [11] chose an unexpanded random node as a core for the expansion. Lee et al. [12] chose a maximum clique as a core. Our choice is motivated by the observation that, overlapping nodes fail to keep stable membership, when an non-determined disjoint community detection is applied. The results of an non-determined disjoint community detection algorithm depend on the ordering of nodes to cluster.

We assume that the probability p_{ij} of a pair of nodes (i, j) belonging to the same community implies their memberships: *(i)* $p_{ij} \geq \theta$ represents (i, j) holding a stable membership (typically $\theta = 80\%$); and *(ii)* if $(i, j) \in C$ and $(i, k) \in C$, then $(j, k) \in C$. Strong clusters are a set of nodes connected by edges $\{i, j\}$ with $p_{ij} \geq \theta$. Under the assumption, a matrix $\mathbf{P} = [p_{ij}]_{n \times n}$ is applied to find strong clusters. We compute the matrix $\mathbf{P}$ by repeating Louvain algorithm [4] until $\|P_{ij}^{k+1} - P_{ij}^k\| < \varepsilon$, where P_{ij}^k represents the results after k runs. An example is shown in Fig. 1. We note the nodes (e.x. $\{v_1, v_2, v_3, v_4, v_5\}$ or $\{v_6, v_7, v_8, v_9, v_{10}\}$), which construct a strong cluster, are assigned to the same community in the ground truth.

Our studies show that the value of the threshold parameter θ affects the stability of results. As known, most of algorithms suffer from the instability of results, such as **LFM** [11] and Louvan algorithm [4]. Too high value of θ results the low stability of

results, that is, different covers can be found on the same graph with the same parameters. But too low value of θ reduces the accuracy of our method. Thus, we tune θ in
$[0.7, 1.0]$ and select the value corresponding to the most stable results.

3.4   Outline of CPC Method

Having reviewing the basic concepts and techniques of our method, we provide an
summery of **CPC** method. Given a graph $G(V, E)$, a scaling parameter α and a threshold
parameter θ, our algorithm is decomposed into three major steps:

1. Detect strong clusters which are composed of strongly connected nodes;
2. Select a core which has non-expanded nodes to greedy expansion;
3. Continues to loop back step 2 until that all nodes have been assigned into at least
   one community.

   We note the complexity of our method depends on the greedy expansion and the core
identification. The greedy expansion requires to update the fitnesses of neighbor nodes
and its complexity is $\mathcal{O}(m \times c)$, where c is the number of cores to be expanded. To
identify cores, we need compute the matrix **P** by repeating Louvain algorithm. Louvain
algorithm is really efficient and most of the time is consumed by the first step. It is unfortunately not proved analytically, but on average it seems that the time complexity is
$\mathcal{O}(m)$. Let K denote the number of running Louvain algorithm, the matrix computation
is in complexity $\mathcal{O}(m \times K)$. Finally, the total cost is $\mathcal{O}(m \times c + m \times K)$.

4   Applications

In order to demonstrate the validity of **CPC**, we apply it on benchmark graphs and three
real social networks, such as Zachary's Karate club network, the US college football
team network [17], and a dolphin social network [14] in this section.

4.1   Artificial Networks

Firstly, we apply **CPC** to artificial networks, which are generated by Lancichinetti et
al. [10][1]. For simplicity, we set $\alpha = 1$ in testing benchmark networks.

4.1.1   Benchmark with Overlapping Nodes
We begin our validity by benchmark graphs with overlapping nodes. Through [10], the
benchmark graphs are randomly generated by a set of parameters: N the number of
nodes, $\langle k \rangle$ the average degree of nodes, μ the mixing parameter, $maxk$ the maximum
node degree, $t1$ the minus exponent for the degree sequence, $t2$ the minus exponent
for the community size distribution, $minc$ the minimum community size, on the number of overlapping nodes and om the number of memberships of overlapping nodes.
In randomly generated graphs, the internal degree of each node $k_i^{in} = (1 - \mu)k_i$ is determined by the value of the parameter μ. The network fuzziness increases with the

---

[1] The source code is available:
   http://sites.google.com/site/santofortunato/inthepress2

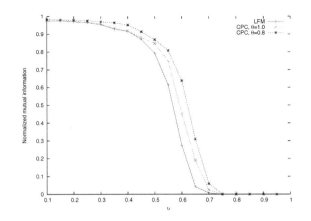

Fig. 2. In increasing of μ, performances of **CPC** and **LFM** on Lancichinetti et al. benchmark networks with $N = 1000, \langle k \rangle = 15, maxk = 50, t1 = 2, t2 = 1, minc = 20, maxc = 50, on = 50$ and $om = 2$ in terms of **NMI**

parameter μ. We apply the Normalized Mutual Information (**NMI**) [11] to compare the community structure found by our method, and the real community structure provided by benchmark graphs. This measure qualifies the similarity between two covers. The higher value of **NMI** is, the more similar tow covers are. If two covers are identical, the value of **NMI** is 1.

We show the results in Fig. 2, where each point corresponds to the average **NMI** over 1000 random graphs with the same parameters. It shows that **CPC** has good performances, particularly with $\theta = 0.8$. Comparing to results of **LFM** [11] on the same benchmark graphs, we note the large variances. Although both **LFM** and **CPC** applied the fitness function to the optimization function with the same parameter $\alpha = 1.0$, **CPC** is more accurate. From the change rate of **NMI** for both **CPC** and **LFM**, we note the importance of cores for community detection. An unsuitable core results that the expansion is ended in advance when a community has not been identified and a new expansion starts. It explains the performances of **CPC** with cores for $\theta = 1.0$ are less accurate than cores for $\theta = 0.8$.

4.1.2   Benchmark of Disjoint Community Structure

Benchmark graphs with disjoint community structure are another good choices to validate **CPC**. A famous benchmark graph is Girvan and Newman benchmark graph [16], which is composed of 128 nodes, divided into 4 communities each of which contains 32 nodes. Edges are placed independently and randomly with μ, that is the ratio of the external degree of each node to its total degree. By increasing the value of μ, the community structure becomes fuzzy and difficult to identify. We plot the results in Fig. 2, where each point shows the average **NMI** over 1000 random graphs with the same parameters. Comparing the performances of **CPC** and **LFM**, the results of **CPC** are better for $\mu > 0.3$, particularly with $\theta = 0.8$. In such benchmark networks, the results of **CPC** are similar to Louvain algorithm which is a disjoint community detection algorithm.

182   Q. Wang and E. Fleury

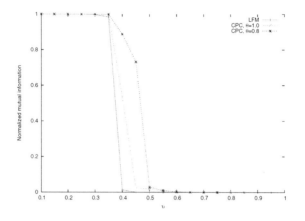

Fig. 3. In increasing μ, performances of performances of **CPC** and **LFM** on Girvan and Newman benchmark graphs in terms of **NMI**

4.2 Real Networks

Considering the heterogeneous of real networks, we apply our method to three real social networks. The accuracy of **CPC** reveals its capability on real networks. In the following, the value of α is gained by running **LFM** in tunning $\alpha \in [0.8, 1.5]$ and the value is selected corresponding to the best results (i.e. the similar number of communities to the ground truth). For the value of θ, it is specified by tunning $\theta \in [0.7, 1.0]$. Since the good performances at $\theta = 0.8$ in artificial benchmark graphs, we select $\theta = 0.8$ to test real benchmark graphs. In fact, the results of **CPC** on the following networks at $\theta = 0.9$ are the same as at $\theta = 0.8$. It indicates that the found peripheral nodes have instable memberships with the core nodes.

4.2.1 Zachary's Karate Club Network
Karate network [26] is a famous social network for testing community detection algorithms. This network characterizes social interactions among members of karate club and it is historically split into two parts. It consists of 34 nodes representing the members of the club and 78 edges showing the friendships between members.

We show the results of **CPC** with $\alpha = 1.0$ and $\theta = 0.8$ in Fig. 4. The overlapping nodes are shown by light green. We observe these overlapping node $9, 10, 31$ are shared by two communities in blue or violet. From the topology, these nodes connect to both communities with the similar number of edges. Comparing to the results in [27], the results of **CPC** is almost totally matched. It demonstrates the ability of our method in community detection. In terms of modularity Q (Eq. 2), it is $Q = 0.419995$ for **CPC** which is similar to $Q = 0.418803$ for Louvain algorithm or $Q = 0.417899$ for **LFM** (Note that the highest value of Q does not represent the best resolution of community structure). It shows that **CPC** has the similar results to Louvain algorithm and **LFM**.

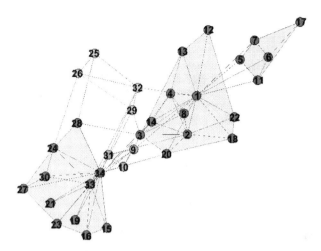

Fig. 4. The overlapping community structure for Zachary's karate club network. The results are gained by **CPC** with $\alpha = 1.0, \theta = 0.8$. Overlapping nodes are node $9, 10, 31$ in light green.

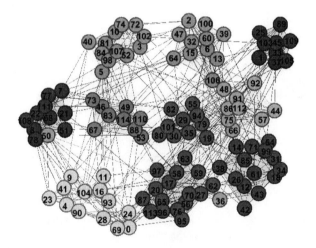

Fig. 5. The overlapping community structure for college football detected by **CPC**. The overlapping nodes are $36, 50$ in cyan.

4.2.2 College Football

We turn to the college football which displays the schedule of division games for 2000 seasons: 115 nodes represent different teams and 613 edges represents the matches occurring between them. These teams construct conference groups, each of which has 8-12 teams and each team plays nearly 7 inter-conference games and 4 inter-conference games. This schedule makes the college football becomes a famous benchmark graph which assigns teams corresponding to their conference groups.

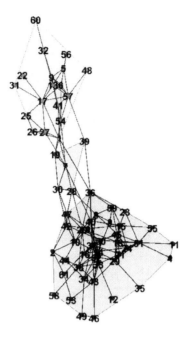

Fig. 6. The overlapping community structure for the dolphin social network. Overlapping nodes are node $7, 19, 28, 30, 39$ in red. These overlapping nodes are covered by green and blue communities.

Applied our method on this network with $\alpha = 0.9, \theta = 0.8$, we show the results in Fig. 5. The found community structure is similar to the results of Louvain algorithm. The overlapping nodes are node $36, 50$, which is also identified in [27]. In fact, the node 36("CentralFlorida") is one of few independent teams. And the node 50("Idaho") belongs to the conference 'Sunbelt'. The conference 'Sunbelt' is one special conference in the network because it plays nearly as many games against 'Western Athletic' teams as they did against teams in their own conference. It explains the assignment of nodes by our method.

The modularity of **CPC** is $Q = 0.611209$, which is similar to $Q = 0.604407$ of Louvain algorithm or $Q = 0.608365$ of **LFM**.

4.2.3   Dolphin Social Network

At last, we analyze the dolphin social network, which is constructed from observations of 62 bottle-nose dolphins over 7 years. Nodes in the network represent the dolphins and edges represent the occurrences of associations between dolphins. Since [17] divided it into 2 clusters and Louvain algorithm uncovers $4 \sim 5$ communities, we applied the parametric modularity [20]. With the parametric resolution $\gamma = 0.4$, Louvain algorithm detects 2 communities. We show the results of **CPC** with $\gamma = 0.4, \alpha = 0.8, \theta = 0.8$ in Fig. 6. Comparing to the results in [15], our method identifies communities approximately exactly. The found overlapping nodes are shown in red, which are node $7, 19, 28, 30, 39$. From the topology, these nodes are border nodes and connect evenly to both communities. The found overlapping community structure are similar to [25]. It demonstrates the validity of our method. In terms of modularity, it is $Q = 0.365798$ of **CPC**, which is similar to $Q = 0.384775$ for Louvain algorithm or $Q = 0.36638$ of **LFM**.

5 Conclusion

In this paper, we have proposed an approach to detect overlapping community structure. Taking benefits of Louvain algorithm, our method detects strong clusters to cores for local expansions. Our analysis on synthetic networks demonstrate that our method has excellent performances in community detection. Especially on benchmark graphs, the results of **CPC** are much better than **LFM**.

Several researches remain. We are currently studying underlying network organizations in both static and dynamic points of view. We are also applying the overlapping structure information to graph layout heuristics inside powerful tools like TULIP[2]. Finally we are investigating the evolution of dynamical communities to mine more structural properties of dynamic networks.

References

1. White, S.: A spectral clustering approach to finding communities in graphs. In: SDM, pp. 43–55 (2005)
2. Barabasi, A.L., Jeong, H., Neda, Z., Ravasz, E., Schubert, A., Vicsek, T.: Evolution of the social network of scientific collaborations. Physica a-Statistical Mechanics and Its Applications 311(3-4), 590–614 (2002)
3. Baumes, J., Goldberg, M., Magdon-Ismail, M.: Efficient identification of overlapping communities. In: Kantor, P., Muresan, G., Roberts, F., Zeng, D.D., Wang, F.-Y., Chen, H., Merkle, R.C. (eds.) ISI 2005. LNCS, vol. 3495, pp. 27–36. Springer, Heidelberg (2005)
4. Blondel, V.D., Guillaume, J.L., Lambiotte, R., Lefebvre, E.: Fast unfolding of communities in large networks. Journal of Statistical Mechanics-Theory and Experiment (2008)
5. Clauset, A., Newman, M.E.J., Moore, C.: Finding community structure in very large networks. Physical Review E 70(6) (2004)
6. Derenyi, I., Palla, G., Vicsek, T.: Clique percolation in random networks. Physical Review Letters 94(16) (2005)
7. Flake, G.W., Lawrence, S., Giles, C.L., Coetzee, F.M.: Self-organization and identification of web communities. Computer 35(3) (2002)
8. Fortunato, S., Barthelemy, M.: Resolution limit in community detection. Proceedings of the National Academy of Sciences of the United States of America 104(1), 36–41 (2007)
9. Hartwell, L.H., Hopfield, J.J., Leibler, S., Murray, A.W.: From molecular to modular cell biology. Nature 402(6761), 47 (1999)
10. Lancichinetti, A., Fortunato, S.: Benchmarks for testing community detection algorithms on directed and weighted graphs with overlapping communities. ArXiv e-prints (2009)
11. Lancichinetti, A., Fortunato, S., Kertesz, J.: Detecting the overlapping and hierarchical community structure in complex networks. New Journal of Physics 11 (2009)

---

[2] http://tulip.labri.fr/

12. Lee, C., Reid, F., McDaid, A., Hurley, N.: Detecting highly overlapping community structure by greedy clique expansion. ArXiv e-prints (February 2010)
13. Li, X., Liu, B., Yu, P.S.: Discovering overlapping communities of named entities. In: Fürnkranz, J., Scheffer, T., Spiliopoulou, M. (eds.) PKDD 2006. LNCS (LNAI), vol. 4213, pp. 593–600. Springer, Heidelberg (2006)
14. Lusseau, D.: The emergent properties of a dolphin social network. Proc. Biol. Sci. 270 (suppl. 2), 186 (2003)
15. Newman, M.E.J., Girvan, M.: Finding and evaluating community structure in networks. Phys. Rev. E Stat. Nonlin. Soft. Matter. Phys. 69(2 Pt 2), 026113 (2004)
16. Newman, M.E.J., Girvan, M.: Finding and evaluating community structure in networks. Physical Review E 69(2) (2004)
17. Newman, M.E.J., Girvan, M., Doyne Farmer, J.: Optimal design, robustness, and risk aversion. Phys. Rev. Lett. 89(2), 028301 (2002)
18. Palla, G., Derenyi, I., Farkas, I., Vicsek, T.: Uncovering the overlapping community structure of complex networks in nature and society. Nature 435(7043), 814–818 (2005)
19. Radicchi, F., Castellano, C., Cecconi, F., Loreto, V., Parisi, D.: Defining and identifying communities in networks. Proc. Natl. Acad. Sci. USA 101(9), 2658–2663 (2004)
20. Reichardt, J., Bornholdt, S.: Statistical mechanics of community detection. Phys. Rev. E 74(1), 016110 (2006)
21. Sales-Pardo, M., Guimera, R., Moreira, A.A., Amaral, L.A.N.: Extracting the hierarchical organization of complex systems. Proceedings of the National Academy of Sciences of the United States of America 104(47), 18874–18874 (2007)
22. Sawardecker, E.N., Sales-Pardo, M., Amaral, L.A.N.: Detection of node group membership in networks with group overlap. European Physical Journal B 67(3), 277–284 (2009)
23. Schuetz, P., Caflisch, A.: Efficient modularity optimization by multistep greedy algorithm and vertex mover refinement. Physical Review E 77(4) (2008)
24. Vazquez, A., Flammini, A., Maritan, A., Vespignani, A.: Global protein function prediction from protein-protein interaction networks. Nature Biotechnology 21(6), 697–700 (2003)
25. Wang, X.H., Jiao, L.C., Wu, J.S.: Adjusting from disjoint to overlapping community detection of complex networks. Physica a-Statistical Mechanics and Its Applications 388(24), 5045–5056 (2009)
26. Zachary, W.W.: An information flow model for conflict and fission in small groups. Journal of Anthropologica 1(33), 452–473 (1977)
27. Zhang, S.H., Wang, R.S., Zhang, X.S.: Identification of overlapping community structure in complex networks using fuzzy c-means clustering. Physica a-Statistical Mechanics and Its Applications 374(1), 483–490 (2007)

Communities Unfolding in Multislice Networks

Vincenza Carchiolo, Alessandro Longheu, Michele Malgeri, and Giuseppe Mangioni

Dipartimento di Ingegneria Elettrica,
Elettronica e Informatica, University of Catania, Italy
{vincenza.carchiolo,alessandro.longheu,
michele.malgeri,giuseppe.mangioni}@diit.unict.it

Abstract. Discovering communities in complex networks helps to understand the behaviour of the network. Some works in this promising research area exist, but communities uncovering in time-dependent and/or multiplex networks has not deeply investigated yet. In this paper, we propose a communities detection approach for multislice networks based on modularity optimization. We first present a method to reduce the network size that still preserves modularity. Then we introduce an algorithm that approximates modularity optimization (as usually adopted) for multislice networks, thus finding communities. The network size reduction allows us to maintain acceptable performances without affecting the effectiveness of the proposed approach.

1 Introduction

Communities structure detection in complex networks is a research field that gained a considerable attention in the last few years. Such interest is due to the possibility to discover hidden behaviours by simply studying the network partitioning into communities. Several methods to address the problem of community uncovering (see [7] for an overview) can be found in literature. However, few of them consider the more general case of communities in time-dependent networks ([5][9][16][2][6]) and/or multiplex networks. On the other hand, networks whose topology evolves over time are quite common ([10][13]). In this case, studying the community structure by simply considering the network obtained by adding together all of its snapshots over time can be too simplistic, and it would not permit to investigate about the temporal evolution of communities. To address this problem, recently Mucha et al. [14] presented a general framework to study the community structure of arbitrary multislice networks, i.e. a set of individual networks linked together by the use of inter-slice links. Multislice networks are general enough to be used to model time-varying, multiplex and multiscale networks. To assess the quality of a given partition into communities, Mucha et al. [14] extended the modularity function, originally introduced in [15](Q_{NG}), to be applied to the more general case of multislice networks ($Q_{multislice}$).

A natural way to explore communities structure in multislice networks is by direct optimization of the $Q_{multislice}$ function. Unfortunately, the exact optimization of the Q_{NG} modularity function is an NP-complete problem [4], and the optimization of the $Q_{multislice}$ function presents a similar problem. To deal with this problem, several approximation methods have been developed (see [7] for an overview). Among them,

L. da F. Costa et al. (Eds.): CompleNet 2010, CCIS 116, pp. 187–195, 2011.

the Louvain method devised by Blondel et al. [3] is one of the fastest yet sufficiently accurate algorithm.

In this work we present an algorithm inspired by [3] to discover communities in large multislice networks. The paper is organized as follows. In section 2 we introduce multislice networks discussing about previous works in this topic. Section 3 presents a method to reduce the size of a multislice network while preserving modularity. Section 4 illustrates our algorithm to discover community structure in multislice networks. Finally, in section 5 conclusions and future works are discussed.

2   Communities in Multislice Networks

Real networks often are inherently dynamic, i.e. they change over time. Community structure in such networks cannot be effectively analyzed neither only considering a single time snapshot nor studying a new network obtained by a sort of "sum" of all the variations across time. On the other hand, traditional approaches to community discovering are not generally well suitable to manage multiplex (or multi–layer) networks, where multiple edges between couple of nodes are allowed. *Multiplex* networks model different kind of relations between nodes and can be, alternatively, represented as a superimposition of distinct layers, each of which being the network obtained by considering a single relation.

To address these issues, in [14] the authors proposed a framework to study community structure in multislice networks. A *multislice* is a network composed by a set of network slices linked together by inter–slices links. As an example of such a network, in figure 1 it is reported a network composed by three slices coupled each other by a set of links depicted using dotted lines. Multislice networks can be used in many contexts. For instance, a multiplex network can be simply represented by a multislice network by mapping each layer of the network to a slice. Moreover, a time varying network can be mapped to a multislice network where each slice is a single instant snapshot network.

In [14] the authors also propose a multislice extension of the Newman's modularity function, thus providing a metric to assess the quality of a given partition into communities of a multislice network.

In particular, given a multislice network, the multislice generalization of modularity for unipartite, undirected network slices and couplings is:

$$Q_{multislice} = \frac{1}{2\mu} \sum_{ijsr} \left\{ \left(A_{ijs} - \gamma_s \frac{k_{is}k_{js}}{2m_s} \right) \delta_{sr} + \delta_{ij}C_{jsr} \right\} \delta\left(g_{is}, g_{jr}\right) \quad (1)$$

Where i and j range over all nodes, s and r range over all slices, A_{ijs} is the element of the adjacency matrix of the slice s (intra–slice), C_{jsr} is the link between node j in slice s and node j in slice r (inter–slice coupling), k_{is} (k_{js}) is the degree of node i (j) in slice s, m_s is the number of links in slice s, γ_s is a resolution parameter and μ is a normalization factor.

Equation 1 can be considered as composed by two terms, the first one takes into account the contribution to the modularity given by each slice (it looks like Newman's modularity), whereas the second term is the contribution given by the inter-slices coupling.

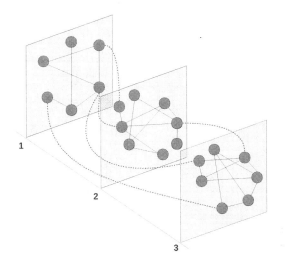

Fig. 1. An example of a three slices network

The modularity function in equation 1 plays a double role: 1) it is used to assess the quality of a given partition and 2) it can be exploited to discover community structure by direct optimization. Unfortunately, as discussed in section 1, the exact modularity optimization is presumably an NP-complete problem (similarly to what has already been observed for Newman's modularity in [4]). To overcome this computability matter, in this paper we propose a greedy method to optimize $Q_{multislice}$ inspired by the Louvain algorithm [3]. In particular, our algorithm makes extensively use of a network size reduction method that we have specifically devised for multislice network, explained in the next section.

3   Size Reduction in Multislice Networks

Reducing the size of multislice networks is useful to implement greedy optimization for $Q_{multislice}$. To achieve this, let G_m a multislice network with undirected network slices and coupling (this does not affect generality).

Note that, by definition, each node in slice s is connected only with the same node in slice r, that is $C_{ijsr} = 0 \ \forall i \neq j$, then $C_{jsr} \equiv C_{ijsr} \ \forall i, j$. This equivalence implies that the term $\delta_{ij}C_{jsr}$ in equation 1 can be substituted by the equivalent C_{ijsr}, so resulting equation is as follows:

$$Q^*_{multislice} = \frac{1}{2\mu} \sum_{ijsr} \left\{ \left(A_{ijs} - \gamma_s \frac{k_{is}k_{js}}{2m_s} \right) \delta_{sr} + C_{ijsr} \right\} \delta\left(g_{is}, g_{jr}\right) \tag{2}$$

Where δ_{ij} has been included into the coupling term.

By definition, for every partition into communities of $G_m \Rightarrow Q^*_{multislice} \equiv Q_{multislice}$.

Now let $Com_s : \{1, ..., N\} \rightarrow \{1, ..M_s\}$ be a partition of slice s of the network into M_s communities. The function Com_s assigns a community index $Com_s(i)$ to node i in slice s of the network G_m. Let us consider the reduced network G'_m obtained as in the following:

- In every slice s we replaced each community with a single node.
- The intra-slice weight w'_{mns} between the nodes m and n of slice s of the reduced network G'_m is defined as in the following:

$$w'_{mns} = \sum_i \sum_j A_{ijs} \delta(Com_s(i), m) \delta(Com_s(j), n) \ m, n \in 1, ..., M_s \tag{3}$$

i.e. w'_{mns} is the sum of all the links connecting vertices in the corresponding communities.

- The inter-slice weight C'_{mnsr} between node m in slice s and node n in slice r of the reduced network G'_m is defined as in the following:

$$C'_{mnsr} = \sum_i \sum_j C_{ijsr} \delta(Com_s(i), m) \delta(Com_r(j), n) \ m \in 1, ..., M_s, \ n \in 1, ..., M_r \tag{4}$$

i.e. C'_{mnsr} is the sum of all the links connecting vertices in community m in slice s with vertices in community n in slice r.

In other words, the reduced multislice network is obtained by collapsing each community in one node and by properly setting the weights of both inter–slice and intra–slice links.

Figure 3 presents an example of the application of the proposed size reduction method. Figure ?? shows the original multislice network G_m composed by three slices where nodes belonging to a community are depicted using the same colour. Figure ?? shows the reduced multislice network G'_m composed (as the original one) of three slices, where each community has been replaced by one node and link weights are recomputed by using equations 3 and 4.

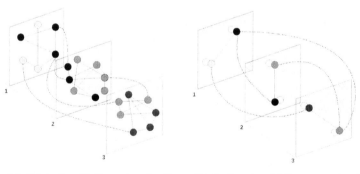

(a) Original multislice network - G_m (b) Reduced multislice network - G'_m

Now we want to prove that the $Q^*_{multislice}$ of G_m is equal to $Q^{*'}_{multislice}$ of G'_m, i.e. the proposed network size reduction method preserves multislice modularity.

The proof that $Q^{*'}_{multislice} = Q^{*}_{multislice}$ is as follow:

$$Q^{*'}_{multislice} = \frac{1}{2\mu'} \sum_{mnsr} \left\{ \left(w'_{mns} - \gamma_s \frac{w'_{ms} w'_{ns}}{2w'_s} \right) \delta_{sr} + C'_{mnsr} \right\} \delta(g_{ms}, g_{nr})$$

$$= \frac{1}{2\mu'} \underbrace{\sum_s \sum_{mn} \left(w'_{mns} - \gamma_s \frac{w'_{ms} w'_{ns}}{2w'_s} \right) \delta(g_{ms}, g_{ns}) +}_{\text{1st term}} \tag{5}$$

$$+ \underbrace{\frac{1}{2\mu'} \sum_{sr} \sum_{mn} C'_{mnsr} \delta(g_{ms}, g_{nr})}_{\text{2nd term}}$$

By applying the same approach followed in [1], it is straightforward to prove that the first term in equation 5 can be rewritten as:

$$1st\ term = \frac{1}{2\mu} \sum_s \sum_{ij} \left(w_{ijs} - \gamma_s \frac{w_{is} w_{js}}{2w_s} \right) \delta(g_{is}, g_{js}) \tag{6}$$

By using equation 4, it is also easy to show that the second term can be rewritten as:

$$2nd\ term = \frac{1}{2\mu'} \sum_{sr} \sum_{mn} C'_{mnsr} \delta(g_{ms}, g_{nr})$$

$$= \frac{1}{2\mu} \sum_{sr} \sum_{mn} \left(\sum_{ij} C_{ijsr} \delta(Com_s(i), m) \delta(Com_r(j), n) \right) \delta(g_{ms}, g_{nr})$$

$$= \frac{1}{2\mu} \sum_{sr} \sum_{ij} C_{ijsr} \sum_{mn} \delta(Com_s(i), m) \delta(Com_r(j), n) \delta(g_{ms}, g_{nr}) \tag{7}$$

$$= \frac{1}{2\mu} \sum_{sr} \sum_{ij} C_{ijsr} \delta\left(g_{Com_s(i)s}, g_{Com_r(j)r} \right)$$

$$= \frac{1}{2\mu} \sum_{sr} \sum_{ij} C_{ijsr} \delta(g_{is}, g_{jr})$$

Putting together the first and second terms, we obtain the following:

$$Q^{*'}_{multislice} = \underbrace{\frac{1}{2\mu} \sum_s \sum_{ij} \left(w_{ijs} - \gamma_s \frac{w_{is} w_{js}}{2w_s} \right) \delta(g_{is}, g_{js})}_{\text{1st term}} + \underbrace{\frac{1}{2\mu} \sum_{sr} \sum_{ij} C_{ijsr} \delta(g_{is}, g_{jr})}_{\text{2nd term}}$$

$$= \frac{1}{2\mu} \sum_{ijsr} \left\{ \left(w_{ijs} - \gamma_s \frac{w_{is} w_{js}}{2w_s} \right) \delta_{sr} + C_{ijsr} \right\} \delta(g_{is}, g_{jr}) \tag{8}$$

$$= Q^{*}_{multislice}$$

In conclusion we proved that nodes belonging to a community in a multislice network can be all replaced by a unique node in the reduced multislice network (this is a generalization of the work by Arenas et al. in [1]).

4   An Algorithm to Discover Communities in Multislice Networks

To discover communities in multislice networks we propose a greedy algorithm based on a local optimization of the modularity function in equation 2.

Given a multislice network G_m , our algorithm consists of two steps repeated iteratively:

Step 1

- Initially, we place each node of the network in a different community, so there are as many communities as the nodes in the multislice network (i.e. $\sum_s N_s$ where N_s is the number of nodes in slice s).
- For each node i in the slice s the gain of $Q^*_{multislice}$ obtained by moving node i in the same community of it's neighbours j is computed. Note that the neighbourhood of a node i is composed by all nodes i is linked to. It also includes those nodes i is linked to through inter-slices coupling.
- Then, node i is placed in the community for which the gain is maximum (and positive).
- Step 1 is performed iteratively until a local maximum of $Q^*_{multislice}$ is reached.

Step 2

After step 1, we build a new multislice network by applying the size reduction method described in section 3.

- Each slice of the new network consists of as many nodes as the number of communities found during the step 1.
- The weight of the intra-slice link between two new nodes i and j is given by the sum of the weights of the links between communities corresponding to nodes i and j respectively (eq. 3). Note that intra-slice links between nodes in the same community are represented by a weighted self–loop link in the corresponding new node.
- The weight C_{ijsr} of the inter-slice links between node i in slice s and node j in slice r is given by the sum of the weights of the links between communities corresponding to nodes i and j placed in slices s and r respectively (eq. 4).

After the second step the number of nodes can diminish drastically, thus speeding up the computation time. To get an idea of how much network size decreases thanks to the proposed reduction method, readers can refer to the work by Arenas et al. [1]. In figure 4 the two steps of our algorithm are graphically illustrated for a network composed by three slices.

In addition, the way the algorithm works permits an implicit discovering of the hierarchical structure of a multislice network. In fact, the network produced at the end of the second step in each pass of the algorithm can be considered as a more higher hierarchical level network. In other words, the hierarchical organization of the network is naturally explored as the algorithm proceeds. In conclusion, our algorithm inherits all the advantages of the Louvain method proposed by V. Blondel et al.[3]:

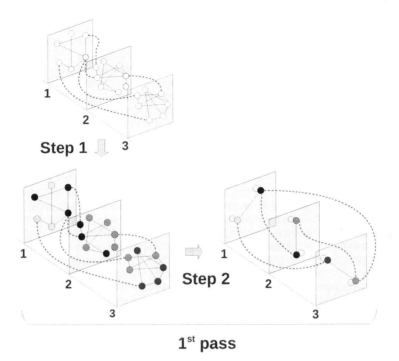

Step 1   3

Step 2

1ˢᵗ pass

Fig. 2.

- It is easy to implement.
- It is very fast (Blondel et al.[3] claim that their implementation is able to find communities in a network of 118 million nodes and 1 billion links in 152 minutes only!)
- It is multi-resolution and naturally gives a hierarchical decomposition of the network.

Additionally, our proposal also allows us to discover communities in multislice networks.

To test our algorithm, we have implemented a prototype written in Python programming language. Since no universally accepted benchmarks for multislice networks currently exist we performed our tests on the examples provided in [14] and specifically we discovered communities in the Zachary Karate network across multiple resolutions. Running our algorithm on this network we obtained the same results reported in [14], thus endorsing the effectiveness of the proposed method.

5   Conclusions

In this paper we presented a greedy algorithm to find communities in multislice networks. Our proposal started from reducing the size of the network without affecting the modularity, so the reduced network partitioning into communities is equivalent to the initial network.

Some issues still have to be addressed, in particular:

- to replace the prototype in Python with an optimized C++ implementation of the greedy algorithm described previously.
- to test the proposed approach to real and large networks, in order to assess its effectiveness as well as its performances. This goal is strictly related to the previous one, indeed an optimized algorithm implementation is essential when working on large networks.
- to investigate about the definition of (new) benchmarks for multislice networks in addition to currently available benchmarks as [11][12][8].

References

1. Arenas, A., Duch, J., Fernandez, A., Gomez, S.: Size reduction of complex networks preserving modularity. New Journal of Physics 9, 176 (2007)
2. Asur, S., Parthasarathy, S., Ucar, D.: An event-based framework for characterizing the evolutionary behavior of interaction graphs. In: Proceedings of the 13th ACM SIGKDD International Conference on Knowledge Discovery and Data Mining, KDD 2007, pp. 913–921. ACM, New York (2007)
3. Blondel, V.D., Guillaume, J.-L., Lambiotte, R., Lefebvre, E.: Fast unfolding of communities in large networks. Journal of Statistical Mechanics: Theory and Experiment (10), 10008 (2008)
4. Brandes, U., Delling, D., Gaertler, M., Gorke, R., Hoefer, M., Nikoloski, Z., Wagner, D.: On modularity clustering. IEEE Transactions on Knowledge and Data Engineering 20(2), 172–188 (2008)
5. Chakrabarti, D., Kumar, R., Tomkins, A.: Evolutionary clustering. In: Proceedings of the 12th ACM SIGKDD International Conference on Knowledge Discovery and Data Mining, KDD 2006, pp. 554–560. ACM, New York (2006)
6. Fenn, D.J., Porter, M.A., McDonald, M., Williams, S., Johnson, N.F., Jones, N.S.: Dynamic communities in multichannel data: An application to the foreign exchange market during the 2007-2008 credit crisis. Chaos 19(3), 033119–+ (2009)
7. Fortunato, S.: Community detection in graphs. Physics Reports 486, 75–174 (2010)
8. Girvan, M., Newman, M.E.J.: Community structure in social and biological networks. Proceedings of the National Academy of Sciences of the United States of America 99(12), 7821–7826 (2002)
9. Hopcroft, J., Khan, O., Kulis, B., Selman, B.: Tracking evolving communities in large linked networks. Proceedings of the National Academy of Sciences 101, 5249–5253 (April 2004)
10. Kumar, R., Novak, J., Tomkins, A.: Structure and evolution of online social networks. In: Proceedings of the 12th ACM SIGKDD International Conference on Knowledge Discovery and Data Mining, KDD 2006, pp. 611–617. ACM, New York (2006)
11. Lancichinetti, A., Fortunato, S.: Benchmarks for testing community detection algorithms on directed and weighted graphs with overlapping communities. Phys. Rev. E 80(1), 016118 (2009)
12. Lancichinetti, A., Fortunato, S., Radicchi, F.: Benchmark graphs for testing community detection algorithms. Phys. Rev. E 78(4), 046110 (2008)

13. Leskovec, J., Lang, K.J., Dasgupta, A., Mahoney, M.W.: Community structure in large networks: Natural cluster sizes and the absence of large well-defined clusters. Internet Mathematics 6(1), 29–123 (2009)
14. Mucha, P.J., Richardson, T., Macon, K., Porter, M.A., Onnela, J.-P.: Community structure in time-dependent, multiscale, and multiplex networks. Science 328(5980), 876–878 (2010)
15. Newman, M.E.J., Girvan, M.: Finding and evaluating community structure in networks. Physical Review E 69, 026113 (2004)
16. Palla, G., Barabsi, A.L., Vicsek, T., Hungary, B.: Quantifying social group evolution. Nature 446 (2007)

Fast Community Detection for Dynamic Complex Networks

Shweta Bansal[1], Sanjukta Bhowmick[2], and Prashant Paymal[2]

[1] Center for Infectious Disease Dynamics, Penn State University,
University Park PA 16802, USA
shweta@sbansal.com
[2] Department of Computer Science, University of Nebraska at Omaha, Omaha NB, USA
sbhowmick@mail.unomaha.edu, ppaymal@unomaha.edu

Abstract. Dynamic complex networks are used to model the evolving relationships between entities in widely varying fields of research such as epidemiology, ecology, sociology, and economics. In the study of complex networks, a network is said to have community structure if it divides naturally into groups of vertices with dense connections within groups and sparser connections between groups. Detecting the evolution of communities within dynamically changing networks is crucial to understanding complex systems. In this paper, we develop a fast community detection algorithm for real-time dynamic network data. Our method takes advantage of community information from previous time steps and thereby improves efficiency while maintaining the quality of community detection. Our experiments on citation-based networks show that the execution time improves as much as 30% (average 13%) over static methods.

1 Introduction

Over the last decade, complex network models have advanced our understanding of systems at all scales, from protein interaction networks to global food webs and from physical transportation networks to online social networks [1,2].

Researchers often build network models from empirical data and then seek to characterize and explain non-trivial structural properties such as heavy-tail degree distributions, clustering, short average path lengths, degree correlations and community structure [3,4,5,6,7]. These structural properties appear in diverse natural and man-made systems, and can fundamentally influence dynamical processes of and on these networks [7,8].

Community structure is a network characteristic describing the propensity of groups of vertices to form dense connections within the group than across. This characteristic is used in the analysis of networks for many applications including hierarchies of organizations [9], collaboration networks [10], protein interactions [1], and stability of electrical grids [11]. The problem of community detection involves finding such connected groups in a given network and has become a popular algorithmic problem in recent years. The quality of a particular division of a network into communities can be measured by modularity. One widely used method of community detection is modularity maximization which detects communities via approximate optimization methods

L. da F. Costa et al. (Eds.): CompleNet 2010, CCIS 116, pp. 196–207, 2011.

such as simulated annealing, spectral clustering, or greedy algorithms, using modularity as the optimization function [12,13].

Much of the current work in complex networks science is based on static networks, which are graphs where vertices and edges remain fixed permanently. However, the diverse interactions that make up empirical complex networks are often quite fluid: new connections form, while others dissolve, providing opportunities for topological changes that can have a major impact on dynamics over the network. Examples include evolving social contact networks over which infectious diseases can disperse and dynamic computer networks over which users can communicate and share resources. A dynamic representation of complex networks, in which vertices and edges shift according to changes in the system, more reflects this reality.

Community detection on dynamic networks has not received much attention until recently. Though consequent temporal configurations of a dynamic network vary by only a small amount (i.e. one node or edge added or deleted), most community detection methods treat each configuration as a separate network. The information regarding communities from the previous configuration is not used and the communities have to be recomputed as a whole, requiring redundant computations. The efficiency of these algorithms can be greatly improved if the re-computation is limited only to the portions of the network that are affected by the modifications.

We propose a fast community detection algorithm for real-time dynamic networks that takes advantage of community information computed in previous time steps and thereby increases the efficiency of the detected community structure. In Section 2, we review the static community detection algorithm on which we base our dynamic algorithm, as well as discuss other approaches to community detection in dynamic graphs. In the next two sections, we discuss our contribution to the dynamic community detection problem, and demonstrate results of our analysis algorithm on a citation-based empirical network. We finish with our conclusions, presenting the benefits of our algorithm and future directions.

2   Background and Previous Work

The idea of communities (or metapopulations in ecology, modules in physics, and cohesive subgroups in sociology) is one that is appealing to many disciplines, and is closely-related to the problem of graph partitioning in graph theory, graph clustering in computer science and block modeling in sociology. In this section, we focus on work in the network science literature based on the hierarchical clustering approach for community detection. For a complete review of other popular methods, see [13].

2.1   Evaluating Community Structure

To develop a method for community identification, one needs an evaluation criteria to judge the quality of the detected community structure. One such measure was proposed by Newman and Girvan in [14] and is based on the intuitive idea that random networks do not exhibit (strong) community structure. Given an arbitrary partition of a network into N_c communities, it is possible to define a matrix C (of size N_c^2) where the elements

C_{ij} represent the fraction of total links starting at a node in group i and ending at a node in group j. Then, the sum of any row of C, $a_i = \sum_j C_{ij}$ corresponds to the fraction of links connected to subgroup i. If the network does not exhibit community structure, or if the partitions are allocated without any regard to the underlying structure, the expected value of the fraction of links within groups can be estimated. It is simply the product of the probability that a link begins at a node in i, a_i, and the probability of links that end at a node in i, a_i. Thus, the expected number of within-community links is a_i^2. The actual fraction of links within each group, however, is C_{ii}. So, a comparison of the actual and expected values, summed over all groups of the partition gives us the deviation of the partition from randomness: $Q(C) = \sum_i(C_{ii} - a_i^2)$. Q is known as modularity, and has become a widely used optimization criteria for community identification algorithms.

The search for the optimal (i.e. largest) modularity value is a NP-hard problem, however, because the space of possible partitions grows faster than any power of the network size. Thus, heuristic search strategies must be used to solve this optimization problem. In recent years, it has been pointed out that the modularity measure has a resolution limit in that it may fail to identify modules smaller than a particular scale (depending on the network size and the degree of module interconnectedness) [15,16]. This phenomenon stems from the fact that modularity is a sum of terms, and thus modularity maximization amounts to searching for the optimal tradeoff between the number of terms (i.e. the number of modules) and the value of each term. Although other quality measures for community structure do exist [17,18], many have the same weakness as the modularity measure because they are calculated as sums over all modules and thus have the same tradeoff as described above.

2.2   Community Detection

Static Community Detection: The hierarchical clustering approach can be dichotomized into divisive and agglomerative strategies, and both require a similarity measure to be defined between vertices or groups of vertices. Agglomerative algorithms [14], begin by initially considering every node in the network to belong to an individual community, and then proceed to combine vertices that are closely related (based on the similarity measure) to form larger communities. Divisive algorithms [12], on the other hand, begin by initially considering all network vertices to belong to a single community and then proceed to remove edges between pairs of vertices that are the least similar. This subdivides the graph into smaller but tighter communities. Hierarchical clustering is a popular community detection method because in addition to providing the identification of communities in the network, it also provides a hierarchical structure in the communities. In, [14], Newman and Girvan, propose a greedy agglomerative approach based on maximization of modularity for hierarchical community identification. In [19], Clauset, Newman and Moore, propose an algorithm (to be referred to as the CNM algorithm from here on forth) which operates on the same principle but is more efficient due to their use of data structures. The algorithm starts from every node in its own community (like all agglomerative approaches), followed by a computation for each pair of communities of the expected increase in modularity if the pair of communities was to be merged. For efficiency, this computation is only made for pairs of communities that are connected, since joining two communities with no connections between them can never result in an increase in Q. The algorithm has running time $O(mdlogn)$ for a network with n vertices, m edges, and a depth, d of the hierarchical community structure, and is thus known to perform efficiently on vertices up to 500,000 vertices [20].

Dynamic Community Detection: The problem of community identification for dynamic network data has received less attention in the field. During the course of the past few years, there have been a few proposed methods, which we review here. These methods fall within two classes: one designed for data which is evolving in real time known as incremental or online community detection; and the other for data where all the changes of the network evolution are known a priori, known as offline community detection.

Tantipathananandh et al [21] propose an offline clustering framework based on (the NP-hard problem of) finding optimal graph colorings. They present heuristic algorithms which find near optimal solutions and are demonstrated on small networks with little evolution. However, these algorithms are likely not scalable in their current form. Ning et al [22] propose an incremental (online) algorithm which is initialized by a standard spectral clustering algorithm, followed by updates of the spectra as the dataset evolves. Compared with re-computation by standard spectral clustering for web blog data, their algorithm achieves similar accuracy but smaller computational costs. More recently, Leung et al [23] discuss the potential of the label propagation algorithm (originally proposed in [24]) for dynamic network data (without any experiments). The label propagation algorithm initializes each node with a unique label, and progresses by allowing each node to adopt the label most popular among its neighbors. This iterative process produces densely connected groups of vertices from the current consensus. In this iterative process densely connected groups of vertices form a consensus on a unique label. The static version of the label propagation algorithm is efficient (near-linear time [24]) and the method could likely be applied to dynamic data. Lastly, earlier in 2010, Mucha et al [25] generalized the Laplacian dynamics approach to obtain a version of the modularity measure for multi-slice (i.e. dynamic) networks. This new measure can then be coupled with existing heuristic methods for dynamic community detection.

3   Our Contribution: Dynamic Community Detection

We introduce a dynamic community detection algorithm for real-time online changes, which involve the addition or deletion of edges in the network. Our algorithm, based on the greedy agglomerative technique of the CNM algorithm, follows a hierarchical clustering approach, where two communities are merged at each step to optimize the increase in modularity of the network. As mentioned in Section 2.1, modularity maximization does have limitations; however, we choose to base our method on the CNM algorithm and the modularity measure because they are well-studied in the field, making comparisons possible.

We observe that if the total number of edges is sufficiently large, then a small change in the number of edges would not significantly affect the fraction of edges in the graph, i.e. the values of C_{ij}. Therefore, until a vertex associated with the modified edge is merged, all earlier merging steps should proceed exactly as in the previous time step. Based on these observations, our dynamic community detection algorithm is designed as follows:

Given a modified edge **(a,b)**, *replicate the combination steps of the previous time steps until vertex* **a** *or vertex* **b** *is encountered. Then switch back to the original agglomerative algorithm and continue as in the static case.*

The primary advantage of our dynamic algorithm is that we reduce redundant computations for identifying communities that are to be merged (and for the merging operation itself) by replicating the merge operations that are common between two consecutive time steps. Depending on the position of the vertices **a** or **b** in the dendrogram representing the hierarchical clustering, our algorithm can reduce as much as 30% of algorithm time compared to a repetition of the static method for each time step. In order to replicate the merges, we store the dendrogram from the previous time step, and this requires $O(n)$ additional memory space, where n is the number of vertices in the network.

We note that our dynamic community detection method is not limited to modularity for the quality function or the CNM technique for hierarchical clustering. We have designed the method so it is conducive to efficient implementation and can be easily added as module to existing agglomerative community detection methods, with any quality functions (local or global).

Pseudocode for Dynamic Community Detection Algorithm
Input: Network G_0 and list of modified edges over time steps where $t = 1, \ldots, T$.
Output: Community structure at time steps $t = 1, \ldots, T$.

1. The community structure of the input network G_0 is initialized using the original greedy agglomerative algorithm.
2. Each combination step is stored as a triplet $< i, j, dQ >, t = 0$, where i and j are communities that have merged and dQ is the increase in modularity due to the merge.
3. For iterations over timesteps $t = 1, \ldots, T$
   (a) Obtain change in edges. Let a and b be the vertices involved in the edge change.
   (b) Update network G_{t-1} to G_t to include the change
   (c) Replicate combination steps of G_{t-1} until vertex a or b is encountered
   (d) Revert to original agglomerative algorithm.
   (e) Continue until increase of modularity, dQ is negative
   (f) Delete combination steps for G_{t-1}
   (g) Store all the combination steps G_t
4. End

Dynamic Updates to Networks: Updating the network structure for each modification is a computationally intensive operation, and whose efficiency depends on the underlying data structure. Data structures for dynamic networks include adjacency lists, such as those used in [26], which are easy to modify through addition and deletion of elements to the list. However, adjacency lists can potentially occupy non-contiguous addresses, thereby not rendering efficient memory utilization.

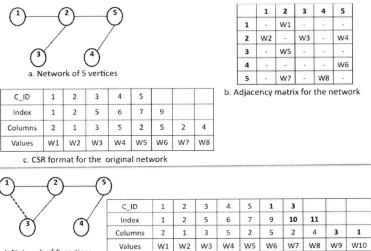

Fig. 1. CSR representation of a dynamic network. Figure a: The original network. Figure b. The sparse adjacency matrix corresponding to the network. The values represent the increase in modularity if the row and column are to be merged. Figure c. The CSR format for the sparse matrix. Figure d. The original network with a new edge (1,3) added, Figure e. Modification to the original sparse matrix to add entries for edge (1,3) and (3,1).

In our implementation, we have used the compressed row storage method [27], a popular format for representing sparse matrices, which stores values associated with the links in the network in an array, i.e. within a contiguous memory location. When an edge is deleted, the corresponding value is set to zero; when an edge is created, a new entry and value is added to the existing array.

CSR is based on expressing the network as sparse matrix, as shown in Figure 1. The array *Index* points to the first non-zero element in each row. The *Columns* array represents the corresponding columns of the matrix with non-zero values and the *Values* array represents the increase in modularity in the communities if the corresponding row and column are joined. To identify the community structure, we add an extra row C_ID to the traditional CSR format which corresponds to the community of each row (vertex). As the agglomeration algorithm progresses the entries in C_ID and *Values* change to reflect the evolving community structure.

The CSR data structure ensures high cache utilization and is easy to implement. However, due to the addition of edges and no deletion (the deleted edges are represented by zeros), the network tends to become larger as the number of modifications increase. In future implementations, we plan to design representations where changes can be consolidated after a certain number of modifications.

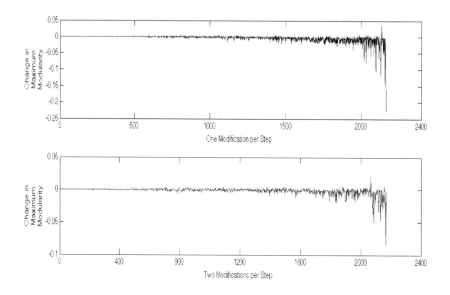

Fig. 2. Difference in maximum modularity of the static and dynamic method over each network snapshot. The X-axis plots the number of modifications and the Y-axis plots the difference in the modularity. Top: One change per time step. Bottom: Two changes per time step.

4   Empirical Results

In this section, we describe the results of our online community detection method on a publicly available empirical network dataset. Although our algorithm is designed for use on real-time data, there are few publicly available datasets of this kind. Thus, to test our algorithm, we use dynamic network data where all temporal snapshots are available a priori, but are processed one step at a time.

The real world network dataset is based on the DBLP database[1] which presents information on computer science publications listed in the DBLP Computer Science Bibliography [28]. The available database provides a snapshot of the bibliography as of April 12, 2006 with article titles, authors, editors, publication dates, venues (journal or conference name) and citation information. From this data, we derive a dynamic co-authorship network spanning the year 2000 to 2001.

The networks have 3252 vertices and from 10997 (for year 2000) to 11159 (for year 2001) edges. Each temporal snapshot network represents authors by vertices and co-authorship by edges. There are 2169 separate changes in the edges (1124 additions and 1044 deletions) from 2000 to 2001. The input to our dynamic algorithm consists of network snapshots given one at a time (to mimic the behavior of a real-time data stream). We compare the time the performance of our dynamic algorithm to that of the

---

[1] Data acquired from http://kdl.cs.umass.edu/data/dblp/dblp-info.html and
http://www.public.asu.edu/ ltang9/heterogeneous_network.html

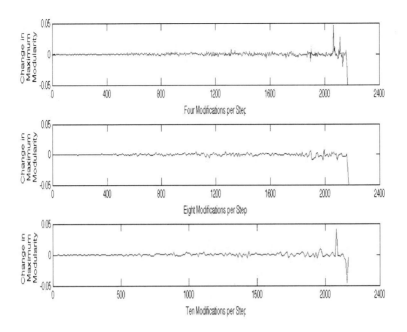

Fig. 3. Difference in maximum modularity of the static and dynamic method over each network snapshot. The X-axis plots the number of modifications and the Y-axis plots the difference in the modularity. Top: Four changes per time step. Middle: Eight changes per time step. Bottom: Ten changes per time step.

static algorithm for the same network at every time step. We experiment with multiple step sizes on 1 change per time step to 2,4,8 and 10 changes per time step. The results and observations of our experiments are described below.

Our dynamic algorithm assumes that small changes in the network will ensure that the total number of edges, and therefore the values of C_{ij}, will remain nearly constant. To fulfill this condition we alternate the online modifications between addition and deletion of edges. However, the total modifications from year 2000 to 2001 in the DBLP dataset, are comprised of 80 more additions than deletions, resulting in a 2% change in the number of edges in the network. As we will discuss below, this change does mildly affect results during the final modifications.

Quality of Solution: To validate the correctness of our algorithm, we compare the solutions of our dynamic algorithm with the analogous static method repeated for every time step. As seen in Figures 2 and 3, the maximum modularity obtained by the two methods remains nearly the same until the final modification steps, where they diverge. The variance in modularity is generally within 5%, except in the case of 1 change per step, where the variance increases to almost 25% during the final modifications. In general, the more changes per step size, the more closely the maximum modularity value from the dynamic method adheres to the static case.

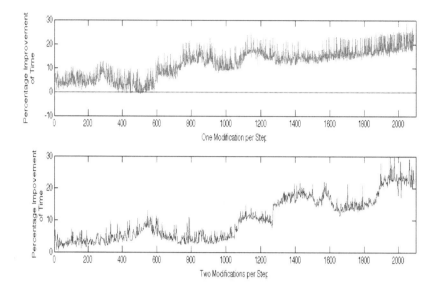

Fig. 4. Percentage speedup of the dynamic method over the static method at each network snapshot. The X-axis plots the number of modifications and the Y-axis plots the speedup. Top: One change per time step. Bottom: Two changes per time step.

The reason for the later discrepancies in the modularity are twofold. First, as discussed earlier, our algorithm is based on the assumption that the total number of edges in the network remain nearly constant. As the number of edges change, the answers tend to diverge. The second reason for the variation in modularity is as follows. The initial merging operations are based on those of the previous time step. However, as we continue adding changes to the network, the early merging steps are determined by the *dynamic* method from the previous steps. The discrepancies from the static algorithm, if any, tend to accumulate to lead to a wider divergence as the number of changes grow. This effect can be ameliorated by reverting to the complete static algorithm after a certain number of network modifications.

Performance Results: Comparing the execution time of community detection for the DBLP dynamic network for the year 2000 to 2001, we find that our dynamic algorithm is either faster or of the same speed as the static case, with efficiency increasing with the number of modifications, which in turn increases the CSR array size. We measure results only for time steps 1 to 2100, since the solutions of the static and dynamic method are equivalent in this range. The percentage of improvement is obtained by computing $\frac{(StaticTime - DynamicTime)}{StaticTime} * 100\%$. As seen from Figures 4 and 5, the speedup can be as much as 30% with an average of 13%. The curves with more changes per time step (Figure 5) are smoother due to fewer calls to the dynamic method, otherwise the progression of the speedup curve is nearly the same over all experiments.

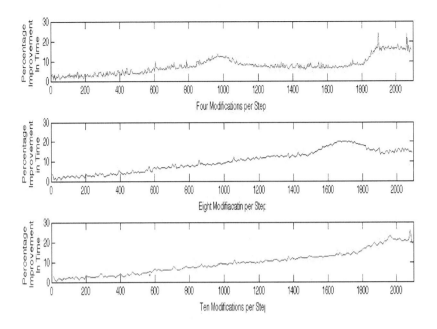

Fig. 5. Percentage speedup of the dynamic method over the static method at each network snapshot. The X-axis plots the number of modifications and the Y-axis plots the speedup. Top: Four changes per time step. Middle: Eight changes per time step. Bottom: Ten changes per time step.

5 Discussion and Future Work

We see from the results, that our real-time dynamic algorithm can improve the execution time of a static agglomerative method, while maintaining the quality of solution as measured by the maximum modularity of the network. However, repeated applications of the dynamic method or too many changes of the same type can hamper the quality of results.

Our goal has primarily been to design an efficient algorithm for dynamic community detection by extending a static agglomerative technique and benchmarking our results with the static algorithm results. We do not claim to improve upon the static algorithm results, nor deal with the issues of the underlying algorithm (as discussed in [15,16]). Finding the "correct" community structure is also an open problem. However, due to the modularity of our algorithm it is easy to add this dynamic component to any current or future agglomerative method that uses hierarchical clustering.

Our future research plans include designing a dynamic component for divisive community detection methods. We also plan to improve the efficiency of the algorithm by a more selective search of the dendrogram and develop an offline-version of our algorithm which we anticipate would perform at least as well as the online version.

Acknowledgements. The authors thank two anonymous reviewers for their insightful feedback. S. Bansal acknowledges support from the RAPIDD program of the Science & Technology Directorate, Department of Homeland Security, and the Fogarty International Center, National Institutes of Health. S. Bhowmick acknowledges the support from the Nebraska EPSCoR First Award and the College of IS&T, University of Nebraska at Omaha.

References

1. Voevodski, K., Teng, S.H., Xia, Y.: Finding local communities in protein networks. BMC Bioinformatics 10(10), 297 (2009)
2. Vazquez, A., Dobrin, R., Sergi, D., Eckmann, J.P., Oltvai, Z.N., Barabási, A.L.: The topological relationship between the large-scale attributes and local interaction patterns of complex networks. PNAS 101, 17940–17945 (2004)
3. Watts, D., Strogatz, S.: Collective dynamics of small world networks. Nature 393(6684) (441), 42–440 (1998)
4. Albert, R., Jeong, H., Barabasi, A.L.: Diameter of the world-wide web. Nature 401, 130–131 (1999)
5. Newman, M., Park, J.: Why social networks are different from other types of networks. Phys. Rev. E 68(036122), 36122 (2003)
6. Newman, M.: Assortative mixing in networks. Phys. Rev. Lett. 89, 208701 (2002)
7. Boguna, M., Pastor-Satorras, R., Vespignani: Epidemic spreading in complex networks with degree correlations. In: Statistical Mechanics of Complex Networks. Lecture Notes in Physics, vol. 625, pp. 127–147 (2003)
8. Albert, R., Barabasi, A.L.: Statistical mechanics of complex networks. Reviews of Modern Physics 74, 47–97 (2002)
9. Porter, M., Mucha, P.J., Newman, M.E.J., Friend, A.J.: Community structure in the united states house of representatives. Physica A 386, 414–438 (2007)
10. Barabasi, A.L., Jeong, H., Ravasz, E., Neda, Z., Schuberts, A., Vicsek, T.: Evolution of the social network of scientific collaborations. Physica A 311, 590–614 (2002)
11. Atkins, K., Chen, J., Anil Kumar, V.S., Marathe, A.: Structure of electrical networks: A graph theory based analysis. International Journal of Critical Infrastructures 5, 265–284 (2009)
12. Girvan, M., Newman, M.: Community structure in social and biological networks. PNAS 99, 7821–7826 (2002)
13. Newman, M.: Detecting community structure in networks. Eur. Phys. J. B 38, 321–330 (2004)
14. Newman, M.E.J., Girvan, M.: Finding and evaluating community structure in networks. Phys. Rev. E 69(2), 026113 (2004)
15. Fortunato, S., Barthlemy, M.: Resolution limit in community detection. PNAS 104(1), 36–41 (2007)
16. Good, B.H., de Montjoye, Y., Clauset, A.: The performance of modularity maximization in practical contexts. Phys. 82, 046106 (2010)
17. Steinhaeuser, K., Chawla, N.V.: Identifying and evaluating community structure in complex networks. Pattern Recognition Letters 31(5), 413–421 (2010)
18. Gaertler, M.: Clustering. Network Anal., 178–215 (2005)
19. Clauset, A., Newman, M.E.J., Moore, C.: Finding community structure in very large networks. Phys. Rev. E 70(6), 66111 (2004)
20. Wakita, K., Tsurumi, T.: Finding community structure in mega-scale social networks. In: Proceedings of the 16th International Conference on World Wide Web, pp. 1275–1276. ACM, New York (2007)

21. Tantipathananandh, C., Berger-Wolf, T., Kempe, D.: A framework for community identification in dynamic social networks. In: Proceedings of the 13th ACM SIGKDD International Conference on Knowledge Discovery and Data Mining, pp. 717–726 (2007)
22. Ning, H., Xu, W., Chi, Y., Gong, Y., Huang, T.: Incremental spectral clustering with application to monitoring of evolving blog communities. In: SIAM Int. Conf. on Data Mining, pp. 261–272 (2007)
23. Leung, I.X.Y., Hui, P., Liò, P., Crowcroft, J.: Towards real-time community detection in large networks. Phys. Rev. E 79, 066107 (2009)
24. Raghavan, U.N., Albert, R., Kumara, S.: Near linear time algorithm to detect community structures in large-scale networks. Phys. Rev. E 76, 036106 (2007)
25. Mucha, P.J., Richardson, T., Macon, K., Porter, M.A., Onnela, J.-P.: Community structure in time-dependent, multiscale, and multiplex networks. Science 328, 876–878 (2010)
26. Bader, D.A., Amos-Binks, A., Chavarrsa-Miranda, D., Hastings, C., Madduri, K., Poulos, S.C.: STINGER: Spatio-Temporal Interaction Networks and Graphs (STING) Extensible Representation, Tech. rep., Georgia Institute of Technology (2009)
27. Saad, Y.: Iterative Methods for Sparse Linear Systems. PWS Publishing Company (1995)
28. The DBLP Computer Science Bibliography, http://dblpVis.uni-trier.de

On Community Detection in Very Large Networks

Alexandre P. Francisco and Arlindo L. Oliveira

INESC-ID / CSE Dept, IST, Tech Univ of Lisbon
Rua Alves Redol 9, 1000-029 Lisboa, PT
{aplf,aml}@inesc-id.pt

Abstract. Community detection or graph clustering is an important problem in the analysis of computer networks, social networks, biological networks and many other natural and artificial networks. These networks are in general very large and, thus, finding hidden structures and functional modules is a very hard task. In this paper we propose new data structures and a new implementation of a well known agglomerative greedy algorithm to find community structure in large networks, the CNM algorithm. The experimental results show that the improved data structures speedup the method by a large factor, for large networks, making it competitive with other state of the art algorithms.

1 Introduction

The problem of graph clustering or community finding has been extensively studied and, for the majority of the interesting formulations, this problem is NP-hard. Thus, in the study of large networks, fast approximation algorithms are required even though we may obtain suboptimal solutions. For a deep review on this topic, we refer the reader to a recent survey on community finding by Fortunato [1]. Here we revisit the modularity maximization problem, which is NP-hard [2], and a well known greedy approach proposed by Newman [3]. The simplest algorithm based on his approach runs in $O(n(n+m))$ time, or $O(n^2)$ for sparse graphs, where n is the number of vertices and m is the number of edges. More recently, Clauset *et al.* [4] exploited some properties of the optimization problem and, by using more sophisticated data structures, they proposed the *CNM algorithm* which runs in $O(md \log n)$ time in the worst case, where d is the depth of the dendrogram that describes the community structure.

In this paper we propose a new implementation of the CNM algorithm, using improved data structures. Although the asymptotic time bound is the same of the CNM algorithm, experimental results show a speed up of at least a factor of two. Moreover, we introduced randomization within our implementation which is useful to evaluate stability as different runs can provide different clusterings. The experimental evaluation includes several public available datasets and benchmarks. We also evaluate the performance of our implementation on large graphs generated with the partial duplication model [5]. The maximum modularity values obtained for these graphs are rather large, which is interesting

L. da F. Costa et al. (Eds.): CompleNet 2010, CCIS 116, pp. 208–216, 2011.

given that these are random graphs. Finally, we briefly discuss the application and integration of this method with other measures and schemata.

2   Algorithm and Data Structures

The proposed algorithm starts with each vertex being the sole member of its community and then, iteratively, it merges pairs of communities that maximize the modularity score Q. Given a graph and a specific division of it into communities, modularity evaluates the difference between the fraction of edges that fall within communities and the expected fraction of edges within communities, if the edges were randomly distributed while respecting vertices degrees [6]. Let $G = (V, E)$ be an undirected graph and A its *adjacency matrix*, i.e., $A_{uv} = 1$ if $(u, v) \in E$, and $A_{uv} = 0$ otherwise. Let n be the number of vertices and m be the number of edges of G. The *degree* d_u of a vertex $u \in V$ is given by $\sum_{v \in V} A_{uv}$. A *clustering* or *partition* $\mathcal{P}$ of G is a collection of sets $\{V_1, \ldots, V_k\}$, with $k \in \mathbb{N}$, such that $V_i \neq \emptyset$, for $1 \leq i \leq k$, $V_i \cap V_j = \emptyset$, for $1 \leq i < j \leq k$, and $\bigcup_{1 \leq i \leq k} V_i = V$. Given a partition $\mathcal{P}$ for G, we compute its *modularity* as

$$Q(\mathcal{P}) = \frac{1}{2m} \sum_{u,v \in V} \left[A_{uv} - \frac{d_u d_v}{2m} \right] \delta_{\mathcal{P}}(u, v), \tag{1}$$

where $m = |E|$ and the $\delta_{\mathcal{P}}$-function is such that $\delta_{\mathcal{P}}(u, v) = 1$ if both $u, v \in C$ for some $C \in \mathcal{P}$, $\delta_{\mathcal{P}}(u, v) = 0$ otherwise. The modularity Q_G of a graph G is defined as the maximum modularity over all possible graph partitions. Although Eq. (1) can take negative values, Q_G takes values between 0 and 1. Values near 1 indicate strong community structure and 0 is obtained for the trivial partition where all nodes belong to the same community. Typically, values for graphs with known community structure are in the range from 0.3 to 0.7 [6,7].

Let $C_i, C_j \in \mathcal{P}_t$ be two communities, where $0 \leq i, j < |\mathcal{P}_t|$ and $\mathcal{P}_t$ is the partition achieved after $t \geq 0$ iterations. The change ΔQ_{ij} in Q after merging C_i and C_j to form a new partition $\mathcal{P}_{t+1}$ is given by manipulation of Eq. (1),

$$\Delta Q_{ij} = Q(\mathcal{P}_{t+1}) - Q(\mathcal{P}_t) = \frac{1}{2m} 2 \sum_{u \in C_i} \sum_{v \in C_j} \left[A_{uv} - \frac{d_u d_v}{2m} \right]. \tag{2}$$

Since calculating the ΔQ_{ij} for each pair C_i, C_j and for each iteration t becomes time-consuming, as in the original CNM algorithm, we store these values for each pair and only update them when needed. Given $C_i \in \mathcal{P}_t$, let $\bar{d}_i = \sum_{u \in C_i} d_u / (2m)$ and assume that $C_i, C_j \in \mathcal{P}_t$ are merged into $C_k \in \mathcal{P}_{t+1}$ at iteration $t+1$. Then, for $\mathcal{P}_{t+1}$, $\bar{d}_k = \bar{d}_i + \bar{d}_j$ and, for each C_ℓ adjacent to C_k,

$$\Delta Q_{k\ell} = \begin{cases} \Delta Q_{i\ell} + \Delta Q_{j\ell} & \text{if } C_\ell \text{ is connected to } C_i \text{ and } C_j, \\ \Delta Q_{i\ell} - 2\bar{d}_j \bar{d}_\ell & \text{if } C_\ell \text{ is connected to } C_i \text{ but not to } C_j, \\ \Delta Q_{j\ell} - 2\bar{d}_i \bar{d}_\ell & \text{if } C_\ell \text{ is connected to } C_j \text{ but not to } C_i. \end{cases} \tag{3}$$

```
struct adj_node {
    int id;
    int u;
    int v;
    struct adj_node *u_nxt;
    struct adj_node *u_prv;
    struct adj_node *v_prv;
    struct adj_node *v_nxt;
};
```

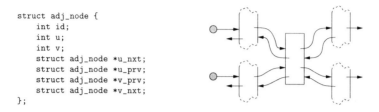

Fig. 1. Cross-linked adjacency list data structure. List nodes are defined by the C structure on the left and are linked as depicted on the right.

These equations follow easily from Eq. (2). Communities are adjacent or connected if there is at least one edge between them. Note that merging two communities for which there is no connecting edge does not increase Q (when C_i is disconnected from C_l, the first term of Eq. (2) is zero and only the second one remains). Therefore, we will not store the value ΔQ for such pairs.

As described above, the algorithm starts with each vertex $u \in V$ being the sole member of a cluster. Let $C_u \in \mathcal{P}_0$ be such that $C_u = \{u\}$, for each $u \in V$. Then, for each $u \in V$ and $(u,v) \in E$, we initially set $\bar{d}_u = d_u/2m$ and $\Delta Q_{uv} = 1/m - 2\bar{d}_u\bar{d}_v$. Accordingly to Eq. (1), the initial value of Q is set to $Q = -\sum_{u \in V} \bar{d}_u \bar{d}_u$. The algorithm proceeds iteratively as follows:

1. select the pair (i,j) with maximum ΔQ_{ij};
2. merge C_i and C_j into C_k (assuming that we are in iteration $t+1$, $\mathcal{P}_{t+1}$ is obtained from $\mathcal{P}_t$ replacing C_i and C_j by C_k);
3. update $\bar{d}_k$ and $\Delta Q_{k\ell}$ for each C_ℓ adjacent to C_k accordingly to Eq. (3);
4. update modularity Q by adding ΔQ_{ij};
5. repeat from step 1 until one community remains.

Here we are assuming that the graph is connected and that we are storing each partition $\mathcal{P}_t$ obtained at iteration t. If the graph is not connected, the algorithm stops when a pair (i,j) does not exist in step 1. Note also that we are usually interested in the partition that maximizes the modularity score. Thus, we can stop when a pair (i,j), selected in step 1, is such that $\Delta Q_{ij} < 0$. By Eq. (3), we know that ΔQ values can only decrease after a such pair be selected and, thus, the modularity value will not increase more since all ΔQ values are negative.

The main point is that we must find the maximum values and extract elements from the adjacency lists as fast as possible. Here we use a single heap data structure to store needed ΔQ values (at most m) and cross-linked adjacency lists to store community adjacencies. Since we have to both decrease and increase values in the heap (recall Eq. (3)), we use a binary heap data structure, for which the get maximum operation takes constant time and the insert, delete and update operations take $O(\log m)$ time, in the worst case. For community adjacencies, we use doubly-linked lists with cross references (see Fig. 1) and, thus, we can solve side effects in constant time when merging two adjacency lists.

Let c_u, c_r, c_t, c_ℓ be real constants. Updating a value in the heap takes $c_u \log m$ time and extracting a value takes $c_r \log m$. Thus, the extraction in step 1 takes $c_r \log m$ time at most and, since there are m elements in the heap, this step is repeated m times at most. Because we get a direct reference in step 1 and we have double-linked lists, removing the edge (i, j) from the community adjacency data structure in step 2 takes constant time. Step 2 requires also $3c_\ell n$ time to merge the adjacencies. Note that there are at most n adjacent communities to C_i and to C_j and that we can solve side effects in constant. Moreover, if a community C_k appears twice in the result, we only keep it once. Thus, to achieve linear time with unsorted lists, without loss of generality, we must process the adjacency of C_i, building a bit array of size n at most, and then process the adjacency of C_j, checking whenever a community C_k occurs in both adjacencies and updating the bit array. Finally we reprocess the adjacency of C_i in order to find the communities C_k which were not in the adjacency of C_j. Step 3 is done along with step 2 and each update takes $c_u \log m$ time at most. Therefore, step 3 takes less than $c_u n \log m$ time. Step 4 takes constant time. Although there exist m elements in the heap, steps 2-4 are executed at most $n - 1$ times and, thus, the running time of the algorithm is at most $c_r m \log m + 3c_\ell n^2 + c_u n^2 \log m$, i.e., $O(n^2 \log n)$ time in the worst case assuming as usual that $m = O(n^2)$.

The differences between our algorithm and the CNM algorithm reside on how we store ΔQ values and how we manage the adjacency of the communities. The CNM algorithm stores ΔQ values in a sparse matrix with each row being stored both as a balanced binary tree and as a binary heap. It maintains also a binary heap containing the largest element of each row. Considering the same max-heap implementation and an efficient implementation of binary trees, updating an element takes $c_u \log n$ and extracting an element takes $c_r \log n$, where n is the maximum size of the heaps in this case. Thus, step 1 takes $c_r \log n$ time. Removing the selected pair from the community adjacency data structure in step 2 takes $2c_t \log n$ to update binary trees plus $2c_u \log n$ to update the heaps. Steps 2 and 3 require also $2n(c_t + 2c_u) \log n + c_u n$ time, since we must update the trees, the k-th heap and the main heap for each C_k in adjacency lists being merged. The heap associated with the resulting adjacency list can be updated in $c_u n$ time. Step 4 takes constant time. Since steps 2–3 are executed at most $n - 1$ times, the running time of the CNM algorithm is at most $(c_r + 2c_t)n \log n + 2(c_t + 2c_u)n^2 \log n + c_u n^2$, i.e., $O(n^2 \log n)$. Note that, although the asymptotic runtime bounds are the same, we get an improvement of at least a factor of two.

For sparse and hierarchical graphs we can provide a better upper bound. A graph $G = (V, E)$ is sparse if $m = O(n)$ and G is hierarchical if the resulting dendrogram for the community merging is balanced. In this case, the sum of the communities degrees at a given depth d is at most $2m$. Therefore the running time is at most $O(md \log n)$, where d is the depth of the dendrogram. Then for sparse and hierarchical graphs, since $m = O(n)$ and $d = O(\log n)$, the algorithm running time becomes $O(n \log^2 n)$. The space requirement of the algorithm is $O(n + m)$ as we store the connections for each community, a total of at most n communities and m connections, and m elements in the heap.

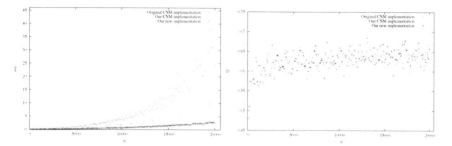

Fig. 2. Average running time and average maximum modularity Q for duplication model graphs obtained with $p = 0.5$. For each n were generated 10 random graphs. The number of edges for those graphs is about 10 times the number of vertices.

We also included in our implementation a randomized edge comparison function. As noted by Brandes *et al.* [2], the algorithm may perform badly if pairs with equal ΔQ are chosen in some crafted order. Although we cannot avoid undesired behavior by ordering these pairs randomly, we expect that it will not happen frequently. With respect to such fluctuations of modularity, we must mention that even small fluctuations may correspond to very different node clusterings [8]. Thus, several runs may be desirable to evaluate the stability of a given clustering, *i.e.*, how stable is vertex assignment along different runs.

3   Experimental Evaluation

In this section we consider 3 implementations in C, the original implementation of the CNM algorithm as provided by the authors, our implementation using optimized data structures to ensure fairness in the comparison and our new implementation. The running times below include the tracking of community membership. For that we use a disjoint sets data structure and, therefore, the running time cost is negligible. All implementations were compiled with the GNU C compiler with flag -O3. The experiments were conduced in a 2.33 GHz quad core processor with 16 GB of memory, running a GNU/Linux distribution.

In order to evaluate the performance on large networks, we generated artificial networks from the partial duplication model [5]. Although the abstraction of real networks captured by the partial duplication model, and other generalizations, is rather simple and no community structure is ensured, the global statistical properties of, for instance, biological networks and their topologies can be well represented by this kind of model [9]. For each number of vertices, we generated 10 random graphs with selection probability $p = 0.5$, which is within the range of interesting selection probabilities [5]. The number of edges for those graphs is approximately 10 times the number of vertices. Fig. 2 provides the running time of our implementation versus the running time of the CNM algorithm, where we observe an improvement of at least a factor of two. We ran also some tests with very large networks and, for a network with 1 million vertices and more than 13 millions edges, our new implementation takes about 9 hours and requires

Table 1. Maximum and minimum modularity for 4 real networks after 1,000 runs. $|V|$ is the number of vertices and $|E|$ is the number of edges for each network. $\max Q$ is the maximum modularity, $\min Q$ is the minimum modularity and $\#\mathcal{P}$ is number of different partitions obtained for 1,000 runs.

Network	$|V|$	$|E|$	$\min Q$	$\max Q$	$\#\mathcal{P}$
Zachary's karate club [10]	34	78	0.381	0.381	1
Bottlenose dolphins'network [11]	62	159	0.492	0.495	2
C. elegans metabolic network [12]	454	2,025	0.385	0.413	253
Protein interaction network [13]	2,215	2,203	0.842	0.846	770

744 MB of memory, while our implementation of the CNM algorithm takes 40 hours and requires 1,796 MB. In Section 4 we discuss how prioritizers can further reduce the running time. Although this model does not ensure any community structure, note that the values of modularity are usually higher than 0.5 (see Fig. 2). This is an interesting fact that deserves a better understanding.

Given that in our algorithm we pick randomly a pair whenever two pairs have the same ΔQ value, we evaluated our implementation on several public datasets and benchmarks, focusing on the stability of the obtained clusterings. Table 1 provides details for four real networks. Unsurprisingly, Q values are identical to those reported by the CNM algorithm. But the partitions found for the last two networks are rather unstable, namely for the protein interaction network where in 1,000 runs 770 different partitions were found. Although we did not analyse further these networks, our results raise an important question concerning partition stability. This is an important issue in the study of networks and, until now, most of the analyses in the literature just consider one partition.

4 Discussion

There are alternative approaches for the greedy optimization of modularity. Schuetz and Caflisch [14] proposed an approach where they merge at once ℓ disjoint pairs of communities instead of just one pair and which can benefit from improvements proposed here. More recently, Blondel et al. [15] proposed an alternative greedy approach. The algorithm proceeds by alternating two main steps. In the first step, it iteratively considers a vertex, removes it from the current cluster computing the change in modularity, and then selects the cluster that provides the better improvement by moving the vertex to that cluster. This is repeated until no change occurs. The second step consists of building a coarsened graph where each cluster becomes a vertex. Then, we iterate these two steps while there are edges in the coarsened graph. Although the authors do not provide a theoretical bound, the running time seems to be almost linear from the experiments, making it one of the fastest algorithms to date. The improvements proposed in this paper make the CNM algorithm competitive with that algorithm, if not faster. With respect to clustering quality, Noack and Rotta [16] stated that the most effective method consists of a multilevel schema, where the greedy approach studied here is used for the coarsening phase, and the first step of the method proposed by Blondel *et al.* for the refinement phase. They used also prioritizers to improve both the running time and the clustering quality.

One of the first prioritizers was proposed by Wakita and Tsurumi [17], favoring the merge of equal size communities, enforcing the running time bound of $O(n \log^2 n)$ for sparse graphs. Since Newman and Girvan [6] proposed the modularity score to account for the intra-cluster density versus the inter-cluster sparsity, given two clusters or communities C_i and C_j, a natural prioritizer is $\Delta Q_{ij}/(d(C_i)d(C_j))$, that favors the merge of clusters with lower weight density, conducting to more dense clusters on average. Although previous studies pointed in this direction [18], only recently Noack and Rotta [16] explicitly used this prioritizer and the variant $\Delta Q_{ij}/\sqrt{d(C_i)d(C_j)}$. Although both prioritizers are related, the second one is closely tied to the null model underlying the modularity measure. By Eq. (2), $2m\Delta Q_{ij}$ is the difference between the observed and the expected number of edges between C_i and C_j. On the other hand, since the null model assumes a binomial distribution, we know that for large graphs the variance of the number of edges between C_i and C_j is approximately $d(C_i)d(C_j)/(2m)$. Thus, the second prioritizer accounts for the number of standard deviations between the observed and the expected number of edges between C_i and C_j. We ran our algorithm with this prioritizer for the network with 13 million edges and we obtained an outstanding speedup, it takes now 130 seconds instead of 9 hours. Also, the values of the modularity did not decrease, as expected, given the modularity definition and its close relation with this prioritizer. As observed before [19,17], the reason for the speedup is that, without any prioritizer, the greedy approach merges the cluster with the largest contribution to modularity with its best neighbor. The strength of the cluster increases and the process continues until all good neighbors are merged. But, since this cluster has great influence, several bad neighbors may also be merged before any other merge pairs be considered. This produces very unbalanced partitions taking the running time up to the upper bound $O(n^2 \log n)$. Note also that the modularity value may be lower since some bad neighbors were merged. In fact, after considering the prioritizer, we obtained higher modularity values. For instance, for the Zachary's karate social network we achieved a value of 0.419. The most important conclusion from our work is that, with the new implementation and considering good prioritizers, we are able to effectively process real large scale-free networks and evaluate its community structure stability.

Several measures have been proposed to evaluate clusterings quality, in particular because modularity suffers some resolution problems [20,21]. Thus, it is important to note that the optimization approach discussed in this paper can be easily adapted for other measures and, in general, it is sufficient to rewrite Eq. (3). This is straightforward for measures based on modularity, such as the modularity for weighted graphs or the similarity-based modularity [22]. For other measures, the updates after each merging step may require some more careful analysis. Nevertheless, we can employ this greedy method either alone or combined with other approaches. For instance, Rosvall and Bergstrom [23,24] used recently this approach in their study of mutual information and of maps of random walks to uncover community structure. In their work they improve the final results by using a simulated annealing based approach [24].

References

1. Fortunato, S.: Community detection in graphs. Physics Reports 486, 75–174 (2010)
2. Brandes, U., et al.: On finding graph clusterings with maximum modularity. In: Brandstädt, A., Kratsch, D., Müller, H. (eds.) WG 2007. LNCS, vol. 4769, pp. 121–132. Springer, Heidelberg (2007)
3. Newman, M.E.J.: Fast algorithm for detecting community structure in networks. Physical Review E 69, 066133 (2004)
4. Clauset, A., Newman, M.E.J., Moore, C.: Finding community structure in very large networks. Physical Review E 70, 066111 (2004)
5. Chung, F., Lu, L., Dewey, T.G., Galas, D.J.: Duplication models for biological networks. Journal of Computational Biology 10(5), 677–687 (2003)
6. Newman, M.E.J., Girvan, M.: Finding and evaluating community structure in networks. Physical Review E 69, 026113 (2004)
7. Guimerà, R., Sales-Pardo, M., Amaral, L.A.N.: Modularity from fluctuations in random graphs and complex networks. Physical Review E 70(2), 025101 (2004)
8. Agarwal, G., Kempe, D.: Modularity-maximizing communities via mathematical programming. The European Physical Journal B 66(3), 409–418 (2008)
9. Bhan, A., Galas, D.J., Dewey, T.G.: A duplication growth model of gene expression networks. Bioinformatics 18(11), 1486–1493 (2002)
10. Zachary, W.W.: An information flow model for conflict and fission in small groups. Journal of Anthropological Research 33, 452–473 (1977)
11. Lusseau, D., et al.: The bottlenose dolphin community of Doubtful Sound features a large proportion of long-lasting associations. Behavioral Ecology and Sociobiology 54(4), 396–405 (2003)
12. Duch, J., Arenas, A.: Community identification using extremal optimization. Physical Review E 72, 027104 (2005)
13. Jeong, H., Mason, S., Barabási, A.L., Oltvai, Z.N.: Centrality and lethality of protein networks. Nature 411(6833), 41–42 (2001)
14. Schuetz, P., Caflisch, A.: Efficient modularity optimization by multistep greedy algorithm and vertex mover refinement. Physical Review E 77(4), 46112 (2008)
15. Blondel, V.D., Guillaume, J.L., Lambiotte, R., Lefebvre, E.: Fast unfolding of communities in large networks. Journal of Statistical Mechanics P10008 (2008)
16. Noack, A., Rotta, R.: Multi-level algorithms for modularity clustering. In: Experimental Algorithms. LNCS, vol. 5526, pp. 257–268. Springer, Heidelberg (2009)
17. Wakita, K., Tsurumi, T.: Finding community structure in mega-scale social networks. In: International World Wide Web Conference, pp. 1275–1276. ACM, New York (2007)
18. Reichardt, J., Bornholdt, S.: Statistical mechanics of community detection. Physical Review E 74, 016110 (2006)
19. Danon, L., Duch, J., Diaz-Guilera, A., Arenas, A.: The effect of size heterogeneity on community identification in complex networks. Journal of Statistical Mechanics P11010 (2006)

20. Fortunato, S., Barthélemy, M.: Resolution limit in community detection. PNAS 104(1), 36–41 (2007)
21. Kumpula, J.M., Saramäki, J., Kaski, K., Kertész, J.: Limited resolution in complex network community detection with Potts model approach. The European Physical Journal B - Condensed Matter and Complex Systems 56(1), 41–45 (2007)
22. Feng, Z., Xu, X., Yuruk, N., Schweiger, T.A.J.: A novel similarity-based modularity function for graph partitioning. In: Song, I.-Y., Eder, J., Nguyen, T.M. (eds.) DaWaK 2007. LNCS, vol. 4654, pp. 385–396. Springer, Heidelberg (2007)
23. Rosvall, M., Bergstrom, C.T.: An information-theoretic framework for resolving community structure in complex networks. PNAS 104(18), 7327 (2007)
24. Rosvall, M., Bergstrom, C.T.: Maps of random walks on complex networks reveal community structure. PNAS 105(4), 111–118 (2008)

Author Index